CORPORATE KILLING

Accident or corporate killing? The Bhopal disaster has been widely seen as an accident whose recurrence could and should be prevented by better safety measures. Yet it can hardly be called an accident, in view of the technological design and management policies that made the poison gas leak almost inevitable.

Official claims reassured us that such a disaster could not occur in an advanced industrialized country. Yet the chemical industry's crisis soon deepened with a similar leak at the Union Carbide Corporation plant at Institute, West Virginia, and again with a toxic fire in Basle.

In this book Tara Jones analyses the crisis management of this relentless 'corporate killing'. The first part looks at how the Bhopal disaster was managed, in the face of determined popular protest. The second part surveys subsequent disasters to document the more general problem of chemical hazards. Jones argues that they arise from an industrial system whose priorities ensure the recurrence of slow Bhopals and mini-Bhopals, if not full-scale disasters.

Author of many articles on the chemical industry, 'Tara Jones' has been involved in campaigns against toxic hazards in Ireland.

Corporate Killing
Bhopals Will Happen

Tara Jones

'an association in which the free development of each
is the condition of the free development of all'

Free Association Books / London / 1988

First published in Great Britain 1988 by
Free Association Books
26 Freegrove Road
London N7 9RQ

Free Association Books would like to acknowledge permission to reproduce extracts from the following publications:
David Halle, *America's Working Man: Work, Home and Politics among Blue-Collar Property Owners*. Copyright © 1984 by University of Chicago Press. Reprinted by permission of the publisher and author.
Charles Perrow, *Normal Accidents: Living with High-Risk Technologies*. Copyright © 1984 by Basic Books, Inc. Reprinted by permission of the publisher.
Health and Safety Executive, list of major hazard sites in Great Britain (Appendix 2), 1986.

© Tara Jones 1988

British Library Cataloguing in Publication Data

Jones, Tara
Corporate Killing: Bhopals will happen.
1. Chemical industry — Accidents —
Safety measures
I. Title
363.1'1966   TP201
ISBN 1-85343-008-0 Hbk
ISBN 1-85343-009-9 Pbk

Phototypeset by Input Typesetting Ltd, London

Printed and bound in Great Britain by
Bookcraft, Midsomer Norton, Avon

do Fhleachta, a bheidh fiche bliain d'aois sa bhliain dhá mhíle is do
Emma a bheidh ceithre bliana déag d'aois: go dtiocfaidís slán

CONTENTS

Acknowledgements   ix
Abbreviations   xi

Introduction   1

Part One: Managing Bhopal

  1 Union Carbide's Response   14
  2 The Indian State's Role   54
  3 The Medical Cover-up   89
  4 The Price of Corporate Killing   121

Part Two: More Bhopals

  5 Fall-out from Bhopal   144
  6 Institute, like Bhopal   163
  7 Fall-out from Institute   187
  8 Accidents will Happen   207
  9 We all Live in Bhopal   247
  10 Against Toxic Capital   271

Appendix 1: Some Toxic Hazards   286
Appendix 2: Major Hazard Sites in Great Britain   289

Notes 295
Bibliography 316
Index 325

Acknowledgements

No book is ever the product of one person's labour, and this book is no exception. Personal thanks are due to Mary, Art, Peter, Izzy and the ever-helpful T. H. C. For help in providing material, I must thank Juliet Merrifield of the Highlander Research Center, New Market, Tennessee; Barbara Dinham of the Transnationals Information Centre, London (who also read and commented on the first draft); Richard Leonard of the US Oil, Chemical and Atomic Workers' Union (OCAW); Tom Ronayne; and Les Levidow of Free Association Books. Les Levidow, my long-suffering editor, is responsible for the existence of this book; and his patience in awaiting its materialization was close to saintly.

I am also indebted to many librarians who allowed me access to their collections, especially the underpaid and overworked staff of the Dublin Business Information Centre and the College of Commerce, Rathmines Library. Thanks are also due to Deena for reading the first half of the book, and to George Bradford and all at the Fifth Estate for helping me maintain my sanity. More than to anyone else my thanks go to Pauline Jackson, who, as well as reading the first draft in new-born baby conditions, provided invaluable support and encouragement.

Abbreviations

CMA	Chemical Manufacturers' Association (USA)
CSIR	Council for Scientific and Industrial Research (India)
EPA	Environmental Protection Agency (USA)
GMBATU	General, Municipal, Boilermakers' and Allied Trades Union (UK)
HSW	*Health & Safety at Work* (UK)
ICEF	International Federation of Chemical, Energy and General Workers' Unions
ICFTU	International Confederation of Free Trade Unions
ICMR	Indian Council on Medical Research
JAMA	*Journal of the American Medical Association* (USA)
JSK	Jana Swaasthya Kendra (People's Health Clinic, Bhopal)
MFC	Medico-Friend Circle (India)
MIC	Methyl isocyanate
OCAW	Oil, Chemical and Atomic Workers' Union (USA)
OSHA	Occupational Safety and Health Administration (USA)
UCC	Union Carbide Corporation (USA)
UCIL	Union Carbide India Ltd
URG	Union Research Group (Bombay)

Introduction

'Until Bhopal the world talked about large-scale chemical disasters as a hypothesis – now they are a reality.' – A senior executive, Chemical Industries Association Ltd (quoted in **Public Relations,** *Autumn 1985, p. 25)*

The leak of toxic gases at Bhopal, India, in 1984 was the worst chemical disaster the world has yet seen. Estimates of those who died in the immediate aftermath of the disaster vary from 2,500 to 10,000 people. Apart from the immediate casualties, it has been reported that two people still die each week in Bhopal from the after-effects of the disaster. The disaster caused not only death but also lingering diseases: tens of thousands of people continue to suffer ill effects and diseases caused by the toxic gases released on the fateful night of 3 December 1984.

Bhopal is the latest in a long line of disasters associated with toxic chemicals: Minamata, Flixborough, Seveso, Michigan State, Love Canal, Ixhuatapec, Cubatão 'and places that never make headlines. The list never ends'.[1] These toxic disasters are certain to continue as the use of toxic chemicals throughout industry increases. New variations on these disasters can also be expected with the coming massive use of biotechnology and biochemicals.

The threat of future Bhopals is in no way confined to peripheral or Third World countries. While the three major toxic accidents in 1984 occurred in Brazil (Cubatão), Mexico (Ixhuatapec) and India (Bhopal), major accidents in process plants occurred in 1985 in the USA at Institute, West Virginia, and in the USSR in Chernobyl; more followed in 1986, in the USA (Gore, Oklahoma) and Switzerland (Basle), and in 1987 in the USA

(Texas City). Thus the hazards associated with toxic chemicals threaten not only the peripheral countries but also the metropolitan heartlands of the USA, Europe and Japan.

This book presents an analysis of toxic chemical 'accidents' from Bhopal to Basle, and the changing response to these 'accidents' by industry and the state. It suggests that the source of these 'accidents' is to be found in the economic and management perspectives that control the design and operation of chemical process plants. Given that these perspectives still control not only chemical process plants, but also other industries which use massive amounts of toxic chemicals, it argues that further toxic disasters on the scale of Bhopal are inevitable. In places as distant as Killala, Co. Mayo (Ireland), Institute, West Virginia (USA), Bombay (India) and Béziers (France), the continued use of toxic chemicals poses the threat of future Bhopals. Nor is this threat simply confined to petrochemical plants: it also includes the transport and storage of these toxic chemicals, and the many other industries that use them.

However, a full evaluation of the threat that toxic chemicals pose cannot stop simply at spectacular outrages such as Bhopal or Seveso. It is necessary also to consider not only 'possible Bhopals', but also 'mini-Bhopals' and 'slow-motion Bhopals'. The former are the smaller leaks of toxic chemicals, which generally result in the deaths of only a handful of people, mainly workers and fire-fighters. The latter include the normal releases of toxic chemicals into air and water, the slow poisoning associated with toxic dumps, and long-term exposure of workers and the general population to low levels of toxic chemicals. Particularly worrying here are the toxic chemicals, such as teratogens, mutagens and transplacental carcinogens, which cross generational barriers, and the widespread poisoning of mothers' milk.

The book's structure is as follows. It begins with a case study of the management of Bhopal as a crisis for both the chemical industry and the Indian state. It describes the medical, legal and political struggles that dogged that management. It examines the effects of Bhopal on the chemical industry world-wide. It then shows how the industry's attempt to limit the resulting crisis to peripheral countries was undermined by the toxic leak at Union Carbide Corporation's plant at Institute, West Virginia, in August 1985. This plant was a show-case of supposedly accident-proof technology

– yet the analysis here reveals the similarities between this leak and the one at Bhopal eight months earlier.

The book continues with an examination of the chemical industry's safety record. It argues that the complexity of chemical process plants, combined with management practices which take workers farther and farther away from any control over the work process, makes yet more accidents inevitable. Such accidents are 'normal' to the industry. It then returns to further case studies of struggles over toxic chemicals and accidents in the USA, South-East Asia and Europe. It ends by drawing some tactical lessons for future … … … presenting, in traditional Irish fashion, … … … roblems that toxic chemicals pose.

Bhopal w… … sis, a period of change and instability … … nong the possible consequences o… … ion Carbide Corporation (UCC), m… … ical industry globally, and a drop in tl… … o peripheral countries for investment… … Bhopal created was one which requi… … nagement. In the management of this… … neglected: the predominant prioritie… … s of UCC and the Indian state. In the… … e dead and the walking wounded we… … ain protagonists acted to minimize dama… … …wn interests.

While Bhopal most obviously created a threat to UCC's continued existence, making it unquestionably a crisis for that company, it also raised problems for the whole range of industries using toxic chemicals. Bhopal presented these industries with a public-relations crisis. They needed to reassure the general public that Bhopal was a rare, chance occurrence that would not be repeated. Closely related to this public-relations crisis was a possible economic crisis, as lowered public confidence in industry's ability or willingness to operate safely could result in further, expensive state regulation of industry.

Bhopal represented a major crisis for that vast sector of capitalist production whose continued profitability depends on untrammelled use of toxic chemicals. Obviously the major sector immediately affected was the petrochemical one, though the agricultural sector's persistent massive use of toxic chemicals also came under immediate attack. The chemical industry had to reassure the general public of the industry's safety and reliability and protect its operations from increased government regulation. The chemical industry had to face the danger that it would become 'the nuclear industry of the eighties'. It had to ensure that 'We all live in Bhopal' did not become a rallying cry in the same way 'We all live in Harrisburg' became a rallying cry after the Three Mile Island nuclear disaster. Bhopal immediately raised public fears regarding the industry's safety to a level never previously reached. The slow poisoning of Seveso and Love Canal paled beside the immediate impact of the toxic chemicals released at Bhopal, which many commentators compared to Hiroshima or Nagasaki.

Thus Bhopal required a reaction from the industry not only in India but globally. How much energy and money it was necessary to expend to manage the crisis varied from country to country. It was in the USA, where opposition to toxic industry was greatest and most consistent, that most energy and money was expended. The US Chemical Manufacturers' Association (CMA) estimated in 1986 that it had spent some quarter of a million dollars on analysing the possible effects of Bhopal. In Europe, by way of contrast, the crisis did not become of much concern until the fire at Sandoz's Basle plant in November 1986. In the peripheral countries, the impact of Bhopal was slighter still, though the burning to the ground of a possibly polluting industrial plant at Phuket in Thailand indicated that the crisis did not pass by the peripheral countries.

The industry's strategy was to limit the crisis and to deny the possibility of Bhopal recurring. Its immediate response was an attempt to limit the crisis to India itself, and if necessary to the peripheral countries in general; at all costs the crisis had to be excluded from the metropolitan countries. Thus the industry argued that standards – of operation, labour and regulation – were higher in metropolitan countries than they were in India; it claimed these higher standards would ensure safety even if toxic leaks did occur. When these reassurances faltered following the almost-identical leak at Institute, it continued to argue that its technology was the best available, but that it was impossible to remove human error. Thus the industry attributed the problem to inefficient humans screwing up superb technology; that is, the technology was faultless, the workers were to blame.

Bhopal also represented a crisis for the Indian state. In Bhopal itself, it was essential that the crisis be limited, that the situation return to normality as soon as possible, that the people's anger be contained, mollified and ultimately repressed, and that financial cost to the state be minimized. But the crisis was more than a public-health crisis and it was not limited to Bhopal. It was also an economic and political crisis for the Indian state. On the political front, the state had to maximize UCC's culpability to minimize its own complicity in the killing. This required the state to move away from its hitherto close connections with UCC, but the move was complicated by the presence of firm UCC loyalists within the state and bureaucratic élites. On the economic front, Bhopal represented a possible obstacle to the high-technology, pro-multinational development policy recently introduced by the Indian Prime Minister Rajiv Gandhi.

Bhopal thus presents us with the opportunity to investigate the management of a chemical and public-health crisis resulting from the unrestrained use of toxic chemicals. It exemplifies the source of toxic catastrophes in economic decisions taken by cost-conscious and loss-fearing companies. Bhopal contains within it most of the major aspects of toxic industry that have been contested separately over the past two decades: lack of information on toxicity, complicity of the state in toxic outrages, long-term effects, the operation of double standards by chemical companies in metropolitan and peripheral countries, the closeness of chemicals for peace to chemicals for war, genetic and reproductive effects on women and men, occupational safety and health and the hazards of pesticides.

Bhopal also allows us to see the various functions that the state, the medical establishment and scientific experts play in managing toxic crises. In their response to a major toxic crisis, the role played by these various bodies is shown more starkly than in their normal functioning. Thus Bhopal and its management exposed the political nature of scientific information and research. The different versions of the causes and results of the toxic gas leak reflect the interests of the various groups that put them forward. In Bhopal it was impossible to maintain that science and technology were neutral, when they were so obviously controlled and manipulated to the benefit of various interests, some of which did not even bother to conceal the strings to which their puppets were attached. Finally, the supposed omniscience of science and technology was once again undermined by the constant squabbling among specialists, while the state and the mass media accepted the experts' baseless reassurances rather than the people's experience and the evidence that was available for all to see.

Background to Bhopal: The Growth of Toxic Capital

The reason for the increase in both chemical disasters and chemical crises in the past two decades is not hard to locate. In the period since the Second World War, the use of highly complex synthetic chemicals has become of crucial importance in industrial production. According to the Organization for Economic Co-operation and Development:

The importance of chemicals in modern societies cannot be denied. The twentieth century has witnessed a chemical revolution which has led to the integration of chemicals into all aspects of everyday life. It is a revolution which has transformed our physical surroundings and has changed our food, our medicines, clothing and construction materials. Indeed it is often difficult to identify manufactured articles which do not incorporate chemicals unknown or in limited use before the turn of the century. More than 50,000 chemicals are in current commercial production and a thousand new chemicals or more are introduced each year. They are produced in volumes which range from less than a tonne up to millions of tonnes per year. In many cases they are essential inputs to production processes in other sectors of industry. (OECD, 1983, p. 13)

The introduction of this chemical technology, with its related work processes, was a great leap forward for capital in terms of productivity, profitability and removing control over work processes from the hands of workers, but it proved to be a leap into the abyss for the rest of us. This massive spread of the use of chemicals has resulted in an increased exposure to toxic chemicals through all stages of what may be described as the toxic cycle (though it is no more a complete cycle than the nuclear cycle). This exposure occurs during mining/extraction, refining, production,

INTRODUCTION 7

Science helps build a new India

Oxen working the fields . . . the eternal river Ganges . . . jeweled elephants on parade. Today these symbols of ancient India exist side by side with a new sight — modern industry. India has developed bold new plans to build its economy and bring the promise of a bright future to its more than 400,000,000 people. But India needs the technical knowledge of the western world. For example, working with Indian engineers and technicians, Union Carbide recently made available its vast scientific resources to help build a major chemicals and plastics plant near Bombay. Throughout the free world, Union Carbide has been actively engaged in building plants for the manufacture of chemicals, plastics, carbons, gases, and metals. The people of Union Carbide welcome the opportunity to use their knowledge and skills in partnership with the citizens of so many great countries.

A HAND IN THINGS TO COME

Process Systems and Services

transportation, consumption and waste disposal. While the production of chemicals has fallen back from its original astounding growth rates, the threat it poses continues to be massive: 'A low-growth scenario of 3.5 percent per annum would lead to a doubling of chemical production over the period 1980–2000, and a potential increase of associated hazards' (OECD, 1983, p. 29).

Thus the toxic nature of its work processes and products has become the most important characteristic of certain sectors of production. Capital earned massive profits in these sectors by ignoring the social costs of toxic production and consumption – costs which were borne not only by workers at the point of production but also by the general population. 'For it is apparent that the "squandering of human lives" does not only occur *within* the gates of the nuclear plant or chemical factory but is as "social" as the labour that produces the radioactive electricity and poisons' (*Midnight Notes*, 1980, p. 23).

In this book, 'toxic capital' refers to that vast sector of capital whose profits are based on squandering human lives. Toxic capital is not confined to the petrochemical sector, just as toxic chemical use is not confined to chemical plants. In 1982 the Alliance for Safety and Health gave some indication of the spread of toxic chemicals:

Within this period, industrial and agricultural production and processes, the production of food additives, fuels, fabrics, construction materials, fertilizers, drugs and pesticides became heavily dependent on highly complex synthetic chemicals. Synthetic textiles and plastics were developed within this period, as was the use of nuclear radiation for power and weapons production. (Alliance for Safety and Health, 1982)

The list of toxic hazards, reprinted (from the same source) in Appendix 1, shows how the toxic effects of these chemicals have spread throughout manufacturing and services.[2]

Capitalism has always advanced by destroying the natural world, but the massive quantitative leap in the production of toxic chemicals represents a still higher stage in this destructiveness. Just as the use of nuclear weapons in war presents capital with the possibility of obliterating life on this planet, so also development by toxification presents major threats to the continued existence of human life. Even some socialists, a group resistant to ecological perspectives, have begun to realize the depth of the toxic problem:

What contemporary socialists must begin to face in the next few crucial years is the very real possibility that a successive build-up of these tragedies and the progressive synthesis of their ill effects may be creating ecological conditions from which no economic system – including socialism – may be able to recover. (George, L. N., 1982, p. 10)

Bhopal: Accident or Corporate Killing?

Bhopal was many different things to different people. For UCC, it was an 'incident'. For Dow Chemical's Douglas Rausch it was an example of a lack of operating discipline. For the mass media, it was the world's worst industrial accident. For other observers it was a tragedy or a disaster. There is no need to indulge in a simple catalogue of horrors. For Robert Engler, 'Bhopal was murder, if not genocide' (Engler, 1985, p. 500).

UCC, a major US-based chemical multinational, exported to Bhopal in India a pesticides manufacturing plant that was defectively designed and dangerously deficient in terms of safety.[3] In connivance with an already venal and corrupt political establishment, it operated the factory in gross disregard of the safety of its workers and the local population. Despite various warning signs, both UCC and the Indian state ignored the threat the plant posed, and the company broke trade-union attempts to organize on safety issues. In the end those who lost were, as always, the poor and oppressed. Meanwhile the state and capital squabble over body counts and compensation, and a few scapegoats[4] are sacrificed at the altar of the people's anger.

Yet none of this was an accident.

It was no accident that such large quantities of toxic chemicals were stored at Bhopal: 'UCC insisted on a process design requiring large MIC [methyl isocyanate] storage tanks over the objections of UCIL [Union Carbide India Ltd] engineers' (ICFTU, 1985, p. 6). It was not an accident that workers were unable to realize a gas leak was about to occur:

Inadequate maintenance was a long-standing complaint at the Bhopal plant . . . Broken gauges made it hard for MIC operators to understand what

The flare tower from where the gas leaked. The tower had been shut down for maintenance a few weeks before the accident. A corroded section of piping was to be replaced, but this had not been done, despite the fact that the work involved could easily have been done in the plant. The purpose of the tower was to burn escaping gases.

was happening. In particular, the pressure indicator/control, temperature indicator and the level indicator for the MIC storage tanks had been malfunctioning for more than a year. (ICFTU, 1985, p. 9)

Nor was it an accident that workers did not succeed in stopping the gas leak: 'It is true that some workers ran from the area during the release. That is what they had always been told to do by UCIL management' (ICFTU, 1985, p. 12). There is no way the workers could have prevented the leak: 'What is certain is why they [the Bhopal victims] died. The design of the safety equipment on Union Carbide's pesticide plant was inadequate' (MacKenzie, 1986, p. 42). Nor was this an accident: 'Few new systems have replaced old safety systems because the task is extremely expensive' (p. 44).

In the interests of safety, the Bhopal workers had been instructed to run against the wind in the event of a gas leak. This is what they ran against. Upwind from the flare tower – the origin of the leak into the atmosphere – there is a concrete wall topped with barbed wire. There were no open gates on that side of the plant.

It was no accident that people were ignorant of the fatal effects of MIC: chemical companies are not noted for warning either workers or the general population of the dangers of their products.

UCIL never provided complete information about these chemicals to workers, government authorities, or community residents. Most of the workers we spoke with said they received no training or information about the hazards of MIC or other toxic chemicals in the plant. Residents of J.P. Nager and other neighbourhoods in Bhopal had little idea what UCIL produced; many residents thought the plant was making 'medicine' for crops. (ICFTU, 1985, p. 11)

It was no accident that maintenance was bad and that inexperienced staff were assigned to the MIC unit. 'The workers we interviewed said that employees were often assigned to jobs they were not qualified to do. This practice was also noted by Union Carbide Corporation in its 1982 inspection report. If the workers refused to do the job which they were assigned to on the grounds they were not trained, their salaries were reduced' (p. 10). These maintenance and staffing cuts were the company's response to the fact the Bhopal plant was not making a profit.

Ignorance of a slightly different kind affected plant management. One Bhopal plant manager told the *Financial Times* (10 December 1984, p. 3), 'We didn't know that such a small amount of gas as leaked had the capacity to destroy human lives to this extent. The lethal properties were not known. We thought our safety controls were adequate, so didn't do any community education.'[5]

It is no accident that toxic capital operates in peripheral countries in an even more flagrantly irresponsible and dangerous manner than it does in metropolitan countries: whether or not the cost savings involved are major determinants of where toxic plants are sited, capital has no compunction about taking advantage of whatever it can get away with in terms of environmental damage and workers' safety and health. Nor was the class nature[6] of the killing an accident: the same process of integrating Indian agriculture into the world market, which provided the market for the pesticides that UCIL produced at Bhopal, also displaced the rural people who came to live in the shanty towns across the road from the killer plant (Susman *et al.*, 1983).

Toxic capital's profits exist on a balance-sheet of death. This death may come swiftly, as it came that night in Bhopal. Or it may come slowly, as it comes to workers in toxic capital's factories and waste dumps, the population who inhabit the chemical wasteland near the factories and dumps, and those who are poisoned by the products from which toxic industry derives its profits. None of this is an accident: it is an integral and unavoidable part of toxic development.

PART ONE
MANAGING BHOPAL

1 Union Carbide's Response

> *'Public opinion about chemicals and the chemical industry is not likely to change very much if all we offer are public statements about our good intentions. What the public needs to see are the actions of an industry determined to operate responsibly and regain the public trust.'* – Warren Anderson, chief executive officer, Union Carbide Corporation (UCC), speaking to the US Chemical Industry Institute of Toxicology, March 1984 (**Chemical Marketing Reporter,** *26 March 1984, p. 5*)

Until the public understood toxic chemicals in a factual context, continued Mr Anderson, the chemical industry would never make genuine progress in public policy nor reach its full business potential. The Bhopal killing gave UCC the chance to live up to its chief executive's words. Yet the best UCC could come up with were public statements about its good intentions and a major attempt to control the 'factual context' of the killing. If Bhopal presented toxic industry with an opportunity to react honourably and out of concern for its victims, thus fulfilling the industry's propaganda claims about its social responsibility, UCC failed miserably.

What was most remarkable about UCC's reaction was how well it fitted the old stereotype of the multinational corporation. UCC responded to Bhopal with lies, half-truths, misinformation and publicity gestures; attempted to shift the blame to its employees and its subsidiary company; and overall showed itself more concerned with protecting its continued existence and profitability than with ameliorating the damage it had caused. What, one wondered, had happened to the new, environmentally sensitive

UCC celebrated by *Business Week* (anon., 1976) and *Fortune* (Menzies, 1978)? More to the point, did it ever exist at all?

UCC's reactions to Bhopal were as hard-nosed as any other corporation's would be. Its activities after the killing showed it to have its priorities in the right capitalist order. While it paid immediate and constant attention to managing the crisis in the metropolitan countries, its reactions to the real crisis in Bhopal were mean, self-interested and limited. More than that, its concern to minimize its legal liability for the killing ensured that the killing would continue. It withheld information on the exact nature of the toxic gases released, it denied the possibility of cyanide-like poisoning and it fostered a controversy over the use of an antidote to cyanide poisoning: all this led to the killing's survivors being denied one possibly life-saving treatment. Bhopal stripped all the rhetoric off the new, caring corporation that UCC had proclaimed itself to be; what remained was the same old multinational ogre obsessed with profits.

Of course, even within UCC, there were differing responses to the killing. Warren Anderson told the *New York Times* (19 May 1985, section 3, p. 8), 'If you listen to your lawyers, you would lock yourself up in a room some place. If you listen to the public relations people they would have you answer everything. I would be on every TV programme.' It is also likely that UCC's immediate response included concern for the victims. In the first few days, UCC sent a telex to the Bhopal authorities about possible cyanide poisoning. That this concern did not last long is no major surprise.

While the killing at Bhopal came as a shock to UCC officers, they immediately planned a reaction. As Anderson later recalled, 'I formed and led a crisis management team composed of people from law, finance and public affairs. That was done within hours after the news broke' (*Chemical Marketing Reporter*, 9 September 1986, p. 54). Anderson took on the chief responsibility for dealing with Bhopal, leaving Alec Flamm, UCC's president, to take care of normal company business.

In a situation where UCC's continued existence was under threat, legal advice intended to limit possible damage to UCC would be influential in forming UCC's response. When a position to protect the corporation from attack was formulated, the multinational's hierarchy ensured the position would be obeyed. Thus even UCIL's management, who were branded as responsible for the Bhopal 'incident' by top UCC management in March 1985, kept to the corporate line.

UCC's immediate priority was to stay in business. As the US and European business press ran scare stories questioning whether UCC could survive, and with the major international weekly *Business Week* raising the spectre of UCC seeking protection under Chapter 11 of the US bankruptcy code, UCC's share prices fell. By 17 December 1984, UCC's common stock declined by more than 25 percent, wiping around $890 million off its market value, while UCC bonds fell more than 80 percent in price. The Indian company's shares were also affected: UCIL shares fell from 29 rupees before the killing to 5 rupees by 5 December. On 6 December, almost 10 percent of all outstanding UCC stock had changed hands, with UCC shares being the most heavily traded on the New York Stock Exchange.

UCC's immediate needs were to contain the crisis, to reassure its stockholders that the corporation would survive and, in the all-important blaze of publicity that followed immediately after the killing, to show the company's concern for the victims while denying legal liability for the killing. Though it was necessary to undertake public-relations gestures in India, the major threat to UCC's continued existence lay in the USA. Outside the USA the only part of UCC vulnerable to threat was UCIL, the largest US-owned company in India. But while UCIL's sales ranked it among the top twenty Indian companies, they were chicken-feed to UCC, whose assets of $10 billion and 1983 sales of $9 billion made it the thirty-seventh largest US corporation. In contrast to the threat in India, the legal battle and its fall-out in the USA could endanger the continued existence of the whole corporation.

The threat to the corporation's existence came first from the fall in investor confidence immediately after the disaster and again after the suits had been filed against UCC in the USA. If the company lost the US suits, it faced the prospect of being forced into Chapter 11, with its ownership being transferred to the gas victims – or at least of UCC's losing its shareholders' confidence, with a consequent crash in its market value. Thus UCC recognized from the start that its primary battle lay in the USA, with India remaining only a side-show to the real action. UCC reacted immediately to the financial threat: as early as 4 December, UCC spokesperson Harvey Colbert said UCC had substantial insurance to cover any suits arising from the disaster. UCC quickly held a closed meeting to reassure industry analysts of the company's continued viability. Alas for UCC, analysts were reported to have found the meeting 'unrevealing' and remained unconvinced.

UCC's immediate reaction was almost a knee-jerk: it denied any difference existed between safety measures at Bhopal and at its sister plant in the USA, at Institute, West Virginia. (Interestingly, though this went largely unreported in the media, Institute is primarily a black community.) Jackson Browning, vice-president for health, safety and environmental affairs at UCC, claimed that 'The Indian plant was designed and built by American nationals. As to the standards . . . they are the same. To the best of our knowledge, our employees in India complied with all laws and we are satisfied with the facilities and the operation of them' (*Guardian*, 7 December 1984, p. 11).

This reaction was essential to deal with immediate charges of hazard export and double standards. Thus UCC's first line was that the equipment installed in Bhopal was made in the USA to US specifications, with safety equipment and standards virtually identical in both Bhopal and Institute. This claim was supported by UCIL officials, who said US technicians had played a major part in setting up the Bhopal plant. Later on, this first reaction proved to be a double-edged sword, as it implied that the killing at Bhopal could recur elsewhere. However, it did divert the Western media's immediate coverage from the question of hazard export to the safety of the chemical industry in the developed countries. While the chemical industry didn't welcome this coverage, it was preferable to the many cans of worms that would be opened if the hazards of their operations in peripheral nations were closely examined. It was essential to deny the existence of a general strategy on the part of toxic capital to take advantage of lower (or non-existent) health, safety and environmental standards in the peripheral countries. Thus the US mass media asked, 'Could it happen here?', rather than 'Why did it happen in the Third World?' The industry was also in a better position to reassure the media of the safety of its operations in the metropolitan countries, where the last major catastrophe at a petrochemical plant had occurred at Seveso in Italy in 1976. For a similar catastrophe in the peripheral countries, it was necessary to go back only one month, to the liquid petroleum gas disaster in Mexico.

This denial of double standards also necessitated the closure of the methyl isocyanate (MIC) unit at Institute. Because this position also amounted to an admission of responsibility by UCC, it was soon to be abandoned. By the second week after the disaster, UCC was insisting only that the process safety *standards* were 'identical' and said implementation of the standards 'is left to the local company' (*Chemical Marketing Reporter*, 17 December 1984, p. 3). Nevertheless UCC admitted 'moral responsibility' for the MIC leak and stressed its sorrow that such an event had occured: 'No words can describe the sorrow felt by the employees of the Union Carbide Corporation for the people of Bhopal', Ronald Wishert, UCC vice-president for Federal government relations, told a Congressional committee on 12 December (*New York Times*, 13 December 1984).

Crisis Management as Information Management

Uncritical articles in the business press dealing with UCC's crisis management have continually stressed how UCC followed a strategy of telling the truth at all times and making available such information as it possessed. UCC was not above congratulating itself on this policy: speaking of UCC's 'commitment to open communications', one company release commended UCC's immediate response: 'At the same time, the decision was taken to make Carbide accessible to the media and share whatever information could be secured.' Nothing could be further from the truth. UCC was described as open to the media and the public, while it actually exercised stringent control over what (if any) information it released; this was a marvellous achievement of public relations and media management, a veritable Madison Avenue masterpiece. On 4 December UCC chief executive officer Warren Anderson was saying, 'Every effort will be made to mitigate against the deep sorrow of the people of Bhopal' and that UCC was being 'as upfront as we can' about the killing's causes, despite only sketchy information. Yet UCC backed off from an undertaking that they would probably allow the press to make a tour of the Institute plant. A company spokesman said the refusal to permit examination of the equipment was a policy decision of top management; plant officials threatened to confiscate a camera if a photographer took pictures of the plant from the company driveway.

The general US press quickly found UCC's image of openness to be just that – an image. UCC didn't exert itself giving out hard information on the Bhopal plant and on the events that led up to the killing. Thus a 1982 UCC report on safety problems in Bhopal and the steps taken to correct them was distributed to the US press, but only after it had been scooped by the Indian press. 'And while Carbide faces complex lawsuits on behalf of the Indian victims and obviously must avoid public pronouncements that prejudice its defence, the company has at times refused to provide even the most elementary information or make its technical people available' (*New York Times*, 14 December 1984, p. D2). UCC, for instance, repeatedly refused to give the press a schematic diagram of the MIC system at Bhopal. The difficulties in extracting clear statements from UCC are best expressed by a reporter for the *New York Times*:

In his handling of the press Mr Browning [vice-president for health, safety and environmental affairs, UCC] has maintained a calm, measured, competent composure while declining to divulge technical information about the Bhopal plant. When reporters raised their voices, the Carbide spokesman responded in warm tones, and when reporters asked questions out of turn they have simply been ignored.

Mr Browning has to be seen to be believed. Try this for size:

Reporter Rick Kilmer: *I think you've said the company was not liable to the Bhopal victims?*
Mr Browning: *I didn't say that.*
Kilmer: *Does that mean you are liable?*
Browning: *I didn't say that either.*
Kilmer: *Then what did you say?*
Browning: *Ask me another question.*
Kilmer: *Under what circumstances would you not be liable? (voice rising in frustration)*
Browning calmly declines to respond. (14 December 1984, p. D2)

Chorus of Reassurance

Both capital and state needed to maintain tight control of information to manage Bhopal. Crucial for UCC was the need to manage information relating to the toxicity and nature of the gases that escaped. While this was of primary importance to UCC in its attempts to limit its liability, it was also important for toxic capital in general. Although the media covered the story in depth for only a short period, it had a field day with the gas leak's acute effects. Given this, in an attempt to manage the impression of toxic disaster that the media conveyed, it was essential that any chronic effects should be played down. Thus a whole chorus of reassurance was summoned to lull the population of Bhopal and eventually the world, both by UCC and the Indian government, which also wished to limit the extent of the Bhopal crisis.

UCC expended much effort on reassuring the shareholders of the company's continued viability, but minimal effort to provide information to the victims of the killing and those treating them, a reaction which shows nakedly the priorities of toxic capital. A foretaste of UCC's tactics was provided by the immediate reaction of the local UCIL company doctor. The Bombay-based trade-union research group, the Union Research Group, reports:

The Union Carbide chief medical officer who phoned Hamidia Hospital [the major hospital in Bhopal] well after the patients started arriving described MIC as an 'irritant' and advised treatment with water, whereas the company knew, and informed every operator in the plant, that it was lethal. Even after its toxicity had become only too evident and the deaths were mounting, the company persisted in underestimating its harmful effects. (cf. Agarwal, 1985, p. 9; Everest, 1985, pp. 14–15)

Fifteen days after the killing, J. Mukund, the Bhopal works manager, was still defending his statement that MIC was only a non-fatal irritant: 'It depends on how one looks at it. In its effects it is like tear gas, your eyes start watering. You apply water and you get relief. What I say about fatalities is that we don't know of any fatalities either in our plant or in other Carbide plants due to MIC.'

URG's report is confirmed by the *Financial Times:*

Doctors struggling to contain the effects were bitter about what they considered a lack of help from Union Carbide.

Professor Heeresh Chandra, a leading pathologist for the Home Office at Bhopal's main Hamidia Hospital, said 'Why hasn't Union Carbide come forward and said "This is the gas that leaked, this is the treatment"? Is it not a moral duty to tell us what was used, what is the treatment, what is the prevention? They have not come forward.

'Somebody has to tell us. People out there are not prepared to eat and drink. A company should put it in the newspapers, a big advertisement, on what can be the after-effects.' (8 December 1984)

This failure by UCC to provide information was confirmed by the international trade-union report on Bhopal:

Medical authorities stated that they received little if any information on the diagnosis and treatment of MIC injuries from either UCIL or UCC. A hospital director told us that he finally found out that the chemical was MIC from a newspaper report on the evening of 3 December; the state health director finally received solid information on the chemical from the World Health Organization several days after the accident. Indeed, the first

communication from UCC in the US appears to have been a telex received 5 December, which briefly outlined possible treatment, but did not fully describe the possible toxic effects of MIC. (ICFTU, 1985, p. 11)

Indeed UCC's strategy required that the possible toxic effects of MIC be played down. This approach was assisted by the US specialists that the company flew into Bhopal. William Brown, associate professor of biological sciences at Carnegie Mellon University, Pittsburgh, said respiratory ailments and blindness in those exposed to (unspecified) low levels of MIC would 'go away. A chemical reaction is taking place in which the molecules of isocyanate will be turned over and excreted by the system.' For those exposed to (again unspecified) high levels, Brown admitted that those who endured total whitening of the eye would never recover their sight and those whose lungs were totally covered by the gas would probably die of respiratory failure. It would be difficult to describe any scientist from Carnegie Mellon as totally neutral, given that confidential research for UCC on MIC's toxicity had previously been carried out there (Morehouse and Subramaniam, 1985, p. 41). On 14 December, Dr Hans Weill, a pulmonary specialist at Tulane University, New Orleans, and Dr G. P. Halberg, a clinical ophthalmologist at New York Medical College, said that most of the leak's survivors 'are likely to recover fully and suffer no long-term ill effects' (*New York Times*, 15 December 1984). Dr Thomas Petty of the University of Colorado, who also came to Bhopal at UCC's expense, said that nearly every poisoning victim he had examined was recovering rapidly from exposure to the gas.

UCC's strategy included limiting the symptoms it would accept as caused by MIC to those affecting the eyes and the lungs, the two bodily systems most obviously damaged by the toxic gases. It also needed to limit or ideally deny long-term effects, which it did by claiming that MIC in contact with moisture quickly breaks down into harmless compounds. As well as this:

One of the US doctors denied that MIC is absorbed into the bloodstream, stating that all damage to the heart, liver and central nervous system is caused by lack of oxygen, while the medical director of Union Carbide stated that the chemical cannot reach the liver or uterus. It is not clear what these statements are based on: certainly not on direct observations or tests. (Health & Safety at Work, *May 1985, p. 40*)

Despite this, Tulane University's Weill told the *Journal of the American Medical Association* in April 1985, 'If there are any systemic effects, the only mechanism I know that would account for them is that, if the lung injury is severe enough to produce a substantial degree of hypoxemia (oxygen depletion in the blood), this could produce changes in any organ – liver, kidney or whatever' (*JAMA*, 12 April 1985, p. 2013).

The visits by these doctors were not solely confined to reassurance. According to Morehouse and Subramaniam,

their role appears to have been at least as much a medical intelligence-gathering exercise for Carbide's legal defence . . . We would be greatly surprised if, assuming the litigation proceeds that far, one or more of these doctors did not turn up as expert witnesses for the defense in any one of the various lawsuits brought against Union Carbide as a result of Bhopal, the burden of their testimony being that of trying to minimize the adverse health impact of the Bhopal disaster. (1985, p. 42)[1]

Attempts by these authors to obtain copies of these doctors' reports and recommendations were unsuccessful. UCC's position was boosted by others, including government experts, who joined in the chorus of reassurance. Dr Jeffrey Kaplan of the US government's Centers for Disease Control and Gareth Green of Johns Hopkins University, Baltimore, Maryland, said during the first week in January that survivors of the gas leak were suffering less health effects than first feared, though thousands might have their lives shortened by chronic lung disease and other respiratory problems. Dr Claude Jaeger, the senior regional adviser for the World Health Organization, said on 10 December that survivors faced no risk of paralysis or kidney and liver diseases, and that pregnant women and foetuses would suffer no damage.

Alan Johns, director of the UK-based Royal Commonwealth Society for the Blind, speaking on 13 December after a four-day tour of Bhopal, said, 'No one has been blinded as a result of the MIC gas. What has happened is that, in many cases, the cornea has sustained some damage, but the indications of the worst cases in hospital are that the damage will not be permanent. People are recovering, but they are frightened and they require reassurance and treatment.'

Others played the old game of blame the victim. Daya R. Varma, professor of pharmacology at McGill University, Montreal, visited Bhopal in January 1986. According to Professor Varma, many of the victims are indigent and malnourished and suffered from tuberculosis and other ailments for a long time. 'If they had been healthier to begin with, he suggests, the gas might not have affected them so severely' (*Chemical and Engineering News*, 11 February 1985, p. 39).

At the same time, there were some discordant notes. Professor Heyndrickx, head of the University of Ghent toxicology department in Belgium, said on 5 December that hundreds of people would die of secondary respiratory and neurological effects, while 'it is certain that unborn children will be born with enormous deformities. One cannot be sure about children conceived later by survivors, but there is a significant risk' (*Irish Times*, 5 December 1984). Professor Heyndrickx, an expert adviser to the United Nations, was not flown into Bhopal by UCC or anyone else. Similarly Dr Yves Ellery, of the General Medical Center, Pittsburgh University in the USA, suggested, ' . . . one of the immediate effects of the Bhopal gas could be the death of the corneal cells, resulting in permanent blindness'. No one invited him to Bhopal either.

Free Flow of Information: Non-existent, as Usual

What was most shocking for some people about the Bhopal killing was the apparent lack of information on MIC's toxicity. To quote the Delhi Science Forum (1985), a group of radical Indian scientists, 'There is hardly any authoritative scientific work available on the consequences of exposure to MIC. Despite widespread use of MIC, the world scientific community has done little work on the consequences of exposure to MIC.' The lack of available information on MIC was confirmed on 13 December by Dr Jeffrey Kaplan of the US Centres for Disease Control, Atlanta, Georgia, who was the leader of the team of medical experts invited by the Indian government to advise it on Bhopal. He admitted, 'The basic issue is that we have very little experience with MIC. What is known could be written up in two or three pages. That is the largest part of the problem.' A scientist at the US Chemical Industry Institute of Toxicology told the *Wall Street Journal Europe* (7 December 1984), 'We can't turn up anything at this point on chronic toxicity', while the US National Toxicology Programme said a single Ames test (a simple biological test for cancer-causing potential) in 1983 was negative.

Permitted MIC exposure levels in the USA and Britain are 0.02 parts per million. These exposure levels are based on the controversial claim that a level of exposure exists below which no harm will occur to the majority of those exposed. While these levels are commonly described as 'safe' levels, it is worth noting that the non-governmental organization that sets these limits for work-place exposure, the American Conference of Government and Industrial Hygienists, describes a threshold limit value as 'the level to which it is *believed* that nearly all workers may be repeatedly exposed day after day without adverse effect' (Kinnersly, 1973, p. 115). On the scientific basis for such thresholds, Fishbein (1981, pp. 368-9) notes in relation to cancer-causing agents, 'It is considered most unlikely that the threshold question can be resolved in the near future. Since there is so little data and so many interpretations, the view is widely held that continuing arguments over thresholds are an exercise in futility.'[2]

According to Dr Peter Merriman of the Chemical Industries Association Ltd, 'At 1 ppm in the atmosphere people's eyes will start to water, and by that stage there may already be a high enough concentration to cause serious internal damage.' The only reported test on human beings took place in the Federal Republic of Germany when four healthy volunteers were exposed for brief intervals (1-5 minutes). At 0.4 ppm volunteers neither smelled it nor noticed irritating effects. At 2 ppm, they reported some irritation, with resulting tears. As the concentration increased, symptoms worsened until they became unbearable at 21 ppm. Despite the general lack of information on MIC's toxicity, the US chemical multinational Du Pont said it has calculated that leakage of one gallon of MIC can cause health problems two miles away.

Of some interest here also is the fact that British toxicologists, such as Dr John Henry of the National Poisons Information Centre, had to refer to the effects of First World War chemical weapons to illustrate MIC's effects. Also Britain's Health and Safety Executive was not prepared to provide the prestigious English medical weekly the *Lancet* (hardly a subversive journal) what little information it possessed on the chemical. This refusal was possibly due to the classified nature of the four research establishments that work with MIC in Britain.[3]

One reason for the lack of publicly available information on MIC's toxicity relates to the importance of information for the chemical industry and the secrecy that routinely surrounds the use of toxic chemicals (Frankel, 1981). To quote the Delhi Science Forum, 'The exact process used and much of the information on the hundreds of reactions of MIC with other organic and inorganic compounds are well-guarded secrets of the Union Carbide Corporation (USA).' As *Chemical and Engineering News* reported (11 February 1985, p. 37), 'Union Carbide toxicologists may have the best information on MIC toxicity around, but they're treating it like a trade secret. Although the company has not allowed its information to be published, it is sharing it with the [US] National Toxicological Program and EPA [US Environmental Protection Agency] . . . Carbide considers details of its findings to be proprietary.' James E. Gibson, vice-president of the US Chemical Institute of Toxicology, said in December 1984 that UCC may have tested MIC for long-term health effects but classified its results as 'trade secrets', which are therefore 'not available. That's not unusual.'

The Indian journal *Sunday* went further in a report based on sources close to the Indian Central Bureau of Investigation:

According to reports seized from the research and development centre at the plant at Bhopal as well as documents traced from other offices of the firm, the corporation had conducted a number of experiments on animals and plants, and was well aware of the effects of MIC. According to one investigating official, the firm used to get piecemeal research done at the various laboratories and then put the results together to arrive at an overall conclusion. The Bhopal Research and Development Centre too had been used for finding out some of the effects of MIC on animals and plants. The firm, however, denied any knowledge of long-term or medium-term effects of the toxic gas following the tragedy. (7/13 April 1985, pp. 19–20)

In March 1986 it was revealed that the firm had undertaken tests relating to that definitely long-term effect, carcinogenicity. Jackson Browning, vice-president of health, safety and environmental affairs at UCC, 'says chronic-type, low-level inhalation studies the company conducted in mice and rats in 1980 "found no carcinogenicity" ', according to *Chemical and Engineering News* (3 March 1986, p. 4).

This lack of information on MIC's toxicity is but a reflection of the similar lack of information on many of the chemicals toxic capital routinely uses and constantly introduces into the production process and the general environment. In MIC's case, however, there was another twist. According to the Delhi Science Forum, 'Much of the results of the findings in these areas are also quite likely to have never been published, given the enormous significance such data has for chemical warfare.' This twist was highlighted by reports about the presence of chemical-warfare experts at Bhopal studying MIC's effects. The closeness of chemicals for peace to chemicals for war was highlighted by Indian charges that UCIL's other Bhopal facility, a major research and development centre, was involved in chemical-warfare experimentation.

The fact that Dr Bryan Ballantyne, director of applied toxicology at UCC in South Charleston, West Virginia, had been employed by the British Ministry of Defence until 1979 at Porton Down, the British chemical warfare research centre, where he experimented on animals with hydrogen cyanide, lends support to the tie-in.[4] Ballantyne kept up the connection with hydrogen cyanide. In 1984 he gave a paper on its effects at the US Army's chemical weapons review in Maryland. The Indian government is making certain that such information will not be restricted any longer. Indian scientists are now experimenting with hydrogen cyanide at Gwalior defence laboratories.[5] As usual, these scientists are working on an antidote.

UCC kept the Indian state in the dark as well as the public. Dr Dass, commissioner in charge of relief of gas victims, told Everest (1985, p. 85), 'In the beginning we did get a couple of telex messages, but they were not very helpful, and since then they have not volunteered any information in so far as the management of these patients is concerned.' In March, an Indian government spokesperson told Derek Agnew, reporter for the British Safety Council magazine *Safety and Risk Management*,

What is handicapping us in our health care is the strange attitude of the American company. Very little is known in India about the gas and its long-term effects.

We have reason to believe that Union Carbide has carried out research on animals and we have specifically asked them for the results of this research. They won't give it to us. We have asked them repeatedly, pointing out that meanwhile we just don't know if we are proceeding on the right lines or not and I'm talking about immediate clinical management as well as long-term control.

Their non-response has been amazing. Why are they so indifferent to us and so secretive? All we want is scientific information which we know they must have. Yet they give weak replies saying 'not much is available'. It has trickled back to us from the USA that they do *have this information but they are sitting on it.*

They haven't even given us their manuals. We've been given to understand that the threshold value for MIC is 0.02 ppm, and beyond that it's dangerous to inhale. We don't really know how many in Bhopal inhaled it. For all we know, people living miles away from the factory might be affected. They might think they have a smoker's cough developing – yet it could so easily be something different.

You won't believe it, but with all this suffering, Carbide are more interested in getting information from us than in helping our relief work. Only the other day two doctors from their HQ in Danbury turned up in Bhopal. They never said why they had come nor what they were about. They disappeared into the night.

Other companies maintained solidarity with UCC by not providing information. Dr Dass complained to Agnew, 'Mitsubishi and Bayer are said to have information on MIC, but we are still waiting for any literature they might have.' Agnew says UCC finally provided information to the Indian government 'presumably between 3 March and 15 March', three months after the killing. Even this was unsatisfactory: in October 1985 Dr Dass was still complaining that he had not received adequate information from UCC.

Union Carbide's Other Bhopal Facility

The MIC and other chemical production units were not the sum total of UCC's facilities in Bhopal. On 13 December 1984, the Press Trust of India reported that UCC was engaged in synthesizing new chemicals and testing them on tropical pests in India in collaboration with UCIL. Much of this work was carried out at UCIL's research station in Bhopal. This 200-million-rupee research centre, the biggest in Asia, has five insect-rearing laboratories and a two-hectare experimental farm for testing chemical agents.

On 25 December 1984 a top Indian 'defence' scientist was quoted by the Press Trust of India to be 'concerned about reports that the centre's studies covered the grey area between agricultural research and anti-crop warfare'. He said the Department of Science and Technology had bypassed a top-level committee, set up in 1975 to screen all collaborative research efforts from a 'security' angle, in permitting UCC's research efforts.

We are unlikely to ever find out what, if any, chemical warfare projects UCC undertook at Bhopal. Britain's Channel 4 television programme, broadcast on the first anniversary of the killing, showed burn marks where a large hole had been dug behind the research centre and much paper burnt. It is known, however, that UCC collected germ-plasm of agricultural plants from the north-east of India; this activity is supposed to be entirely under the control of the Indian National Bureau of Plant Genetic Resources.

The Bhopal research centre was undoubtedly important to UCC. While the pesticides plant was de-skilling its work-force, the research centre over the three years before the killing had recruited eight Ph.Ds, thirteen M.Scs, six graduates and three undergraduates. The Delhi Science Forum asked whether, 'given the differential investments in and utilization of the production and research and development facilities . . . the production is not merely a cover for the research and development activities'. This suspicion has led to the most fanciful and inadequate explanation of the killing: that it was a deliberate chemical warfare experiment on the part of the USA. This allegation was mainly made by pro-Soviet elements in India.

> It's worth quoting Anand Grover's response to this charge: 'I think there are much more fundamental reasons, which are continuing, which are responsible. I mean, it's very easy to make that allegation but then you can't explain why a lot of other incidents of that type occurred' (Jackson *et al.*, 1985, p. 13). Fittingly, this theory found its counterpart on the far right, in the California Institute of International Studies, whose Ronald Hilton suggested investigating the 'possibility that Soviet agents were involved'.

Anderson's Flight into India and Other Public-relations Gestures

UCC's 'quick and compassionate' reaction was cited by William Stover (1985) of the US Chemical Manufacturers' Association as one reason why Bhopal did not become more of a crisis for the chemical industry. To complement its tight control of information, UCC recognized it needed a public-relations campaign. This campaign varied from flying flags at half-mast to a minute's silence at the corporation's facilities world-wide: 'the only sound [at the headquarters canteen] was a few people weeping', one UCC employee told *Business Week* (24 December 1984, p. 45).

It was essential to show that UCC was not a cold, uncaring corporation, so UCC was presented as a group of people 'caught between the instincts of human compassion, the demands of public relations [and] the dictates of corporate survival'. For some head-office employees Bhopal was reported to have provoked crises of conscience which led to 'a special intensity . . . in church services here and other religious gatherings, such as the men's prayer breakfast last Saturday at the New Fairfield United Methodist Church five miles north of Danbury', whose pastor avowed, 'these people have a real sense of anguish'. It was touching to see the *New York Times* business pages provide this human interest angle.[6]

The supreme public-relations gesture was Anderson's trip to Bhopal, accompanied by a team of experts. 'It was little more than symbolism, but at times like this symbolism becomes very important', according to John Jeuck, management professor at the University of Chicago School of Business, who compared Anderson's visit to 'those of governors and presidents visiting the scene of disasters in the US'. It was good TV and gave camera people something else to focus on aside from dead bodies.

Anderson's arrest was also good TV. To begin with, we must debunk any notion of Anderson's personal courage in visiting Bhopal. Before heading for India, Anderson had asked Indian officials in the USA for a safe conduct. He was told to contact officials in India's Department of the Environment, who would help with arrangements when he arrived. Anderson didn't bother with the bureaucratic niceties; instead he went to Bhopal 'without telling local authorities'. Therefore Anderson himself was responsible for the breakdown in communication that led to his arrest. There was also the question of Anderson's personal safety in Bhopal to be considered as well as any riot-inducing effects that his walking around Bhopal could produce.

To deal with these problems, the local state answered Anderson's public-relations spectacle with a spectacle of their own. They arrested Anderson and accompanying officials and charged them with a series of serious offences, several of which were non-bailable. Indeed, one early release from an Indian newsagency said the UCC executives could 'be sentenced to death'. A little gentle persuasion from the US embassy was required to prevent Anderson's flight into India from being more permanent than he intended. As though to emphasize the extent to which the state was protecting him, Anderson was flown out of Bhopal by military aircraft. A similar concern for his safety had led to his being protected by fifty armed guards while he was being held in the UCIL guest-house. While he was being held there 100 protesters picketed the building carrying placards calling for Anderson to be hanged.

Nevertheless, in keeping with the sweet, reasonable image, Anderson said he would return to India to stand trial if necessary, but felt the Indian state weren't going to push their luck that far. His arrest and release was denounced as 'a big fraud on the people' by Chandra Shekhar, president of the Janata Party, who said Anderson had been released on 'express instructions of high-ups in the government' who feared prosecution would 'reveal the complicity and negligence of the state and central government'.

The spectacle of Anderson's visit has been well categorized by Robert Engler (1985, p. 496): 'There, in a sequence worthy of Brecht, after his arrest and speedy release on bail, he pledged financial aid and offered to convert the company's modern guest-house, in which he had been detained, into an orphanage.' Better was to come, however. Keeping up the orphanage line, thus playing down the fact that nearly half of those who died were children, UCC flew Mother Teresa into Bhopal. A Catholic nun who won the Nobel Peace Prize in 1979, she had become famous for her work with orphans in Calcutta. Mother Teresa was accompanied in her tour around Bhopal by two government cars and an escort of six armed police in a blue van. On to the gas victims she pressed miraculous medals, 'small aluminium medals of St Mary', which were inscribed 'O Mary, conceived without sin, pray for us who have recourse to thee' (*New York Times*, 12 December 1984, p. A10). She suggested the recipient should 'tie it around your neck or on your arm. It does help if you have faith.' Mother Teresa preached a message of forgiveness: 'This could have been an accident, it is like a fire [that] could break out anywhere. That is why it is important to forgive. Forgiveness offers us a clean heart and people will be a hundred times better after it.' Alas, many of those she visited in hospital did not know her, either by name or by face – not surprisingly, given that most of Bhopal's residents are Muslims and Hindus. Nevertheless, Mother Teresa's message fitted her image as one who would work for a peace which is acceptable to dynamite manufacturers.[7]

UCC's public-relations campaign on Bhopal has been compared with Johnson & Johnson's campaign to save Tylenol, whose market share was threatened when deaths resulted from someone poisoning the drug with cyanide. The similarity was partly due to the fact that public-relations firm Burston Marstellar handled both contracts. The major difference is that UCC's did not work.

At the end of December 1984, when most of the media massaging and reassurance operations were in full swing, a *Business Week*/Harris poll provided disquieting results for UCC. Fully 44 percent of those polled believed UCC had not told the truth about Bhopal, as against 36 percent who believed they had and 28 percent who weren't sure. The poll also found most Americans blamed UCC or UCIL management for the killing, with only 18 percent blaming the Indian workers and 12 percent the Indian government. On a more general level, 49 percent of those polled were convinced US multinationals have lower health and safety standards overseas, as against 38 percent who thought standards were identical (*Business Week*, 31 December 1984, p. 28).[8]

Transfer Blaming

UCC and the Indian state now began a macabre dance around the killing. UCC's first dance was away from its subsidiary company, Union Carbide India Ltd. It suddenly began to refer to UCIL as an 'affiliate', even though it owned a majority of shares in the company. Normally a subsidiary company is defined as 'a company *controlled* by another, the *controlling* company owning over 50 percent (sometimes 100 percent) of the ordinary shares of the subsidiary company' (Hanson, 1977, p. 427, emphasis added), while an affiliate is defined as 'a firm which is associated with another, generally as its subsidiary' (Hanson, 1977, p. 9).

Secondly, UCC began to reverse its initial claim that the plants at Bhopal and Institute were identical. On 4 December 1984, a UCC news release had stated, 'The Bhopal facility is a state-of-the-art plant. It is essentially the same, on a smaller scale, as an Institute, West Virginia, plant – the company's only other methyl isocyanate production facility. The safety precautions for working with methyl isocyanate at both facilities are the same.' By the second week after the killing, this position had changed: now only the safety *standards* at both plants were identical. When a break came in the blanket reassurances from UCC on the high safety standards at Bhopal, it was angled in such a way as to emphasize Indian responsibility. Thus C. S. Tyson, UCC plant inspector, said in mid-December that the Bhopal plant is 'an entirely different set-up' from that at Institute, having been deliberately designed to be more labour-intensive to create more jobs: for instance, safety devices at Institute are automatic, but at Bhopal they must be turned on manually (*New York Times*, 12 December 1984, p. 8). According to Jackson Browning, the reason for non-installation of automatic safety equipment at Bhopal was 'insufficient back-up systems and spare parts in India'. You pays your money and you takes your choice.

This changing of UCC's tune was no doubt influenced by the immense liabilities stemming from Bhopal. Totally reversing their previous position, UCC now claimed it had provided only general safety standards for Bhopal. Jackson Browning said: 'Responsibility for the detailed design is that of the Indian affiliate. My understanding is that the design was executed in India and the choice of pumping and piping origination and things of that sort were made there, all of them designed to respond to the process safety standards we had told them were necessary.'

Browning, UCC's vice-president for safety, health and environmental affairs, admitted that UCC had been 'available for consultation' and that 'a large number of Americans' were on the site and involved in the operation of the plant when it was started up and before it was handed over to UCIL. The same day UCC said it didn't have any detailed plan of the Bhopal facility to hand in the USA. Thus the responsibility for the actual design of the plant, which was already becoming a central issue, moved from the US company to its Indian subsidiary, which was suddenly demoted to the status of a mere 'affiliate' of UCC. This transfer blaming was to be UCC's constant refrain from then on. UCC backed up its version of events by claiming that the design of the plant had to take place in India, owing to Indian law. The transfer blaming was extended to include the victims. Jackson Browning said on 6 December 1984:

The plant's location, three to four miles from the centre of Bhopal, was selected by Union Carbide India Limited more than seventeen years ago. At that time, the area was not densely populated. In India, land is scarce and the population often gravitates towards areas that contain manufacturing facilities. That's how so many people came to be living near the fence surrounding our property. (UCC news release P–0082–84, p. 3)[9]

By 12 December UCC was arguing explicitly that the fault lay in India: 'Union Carbide officials on Tuesday said the plant had been designed in India, an uncommon industry practice. The plant is staffed entirely by Indians. A 1982 Union Carbide audit report and recent reports from India indicate that worker performance was below American standards.' UCC said it had no direct involvement in the construction and design of the Bhopal plant. It contended it had supplied only general process safety standards to UCIL, which alone was responsible for the Bhopal plant's detailed design, equipment selection and plant construction. As though to emphasize their distance from UCIL, UCC officials said they did not even have such basic information as whether the Bhopal plant had installed a water spray recommended by American UCC officials in 1982.

UCC had already shown similar ignorance on 6 December, three days after the killing, when its headquarters responded to reports of accidents and one death at Bhopal since 1981: UCC said it was under the impression that its Indian safety record was good and it was not aware of any fatalities or accidents. The story later became, 'No previous incidents of multiple deaths' (UCC news release, P–0082–84, p. 1); this quiet change of gears was only to be expected from a corporation 'that regards a single death associated with its own or affiliate company operations as a calamity' (UCC) but whose history is littered with corpses.[10]

According to UCC officials, furthermore, the 'frequency of (safety) audits' was decided by UCIL officials rather than by UCC. This of course explained why there had been no safety audit by US officials since 1982. It was reported that elsewhere these audits took place every three years, yet Bhopal was inspected only once in seven years of operation. US experts wisely concurred with UCC and argued that majority ownership of a unit does not have to mean control of that unit. A report filed by UCIL after the 1982 audit said that 'present facilities are adequate'. While UCC officials said many of the concerns raised by the 1982 audit had been corrected, one of the auditors, C. S. Tyson, said he was still concerned over the plant's overall safety planning. By law this planning was vested in the staff of UCIL.

This transfer blaming was struck two severe blows. In an affidavit filed by a former UCC executive who had once headed UCIL, Edward Munoz swore that in the early 1970s, he 'represented the Union Carbide India Ltd position that only token storage [of MIC at Bhopal] was necessary, preferably in small containers, based on both economic and safety considerations'. However, a UCC corporate engineering group 'imposed the view and ultimately made to be built large bulk storage tanks patterned on the similar UCC facilities at Institute, West Virginia' (*Wall Street Journal Europe*, 4 February 1985, p. 2).

This storage decision resulted from economic motives. Dr Varadarajan, of the Indian Council for Scientific and Industrial Research, said in January 1985 that UCIL had 'many times its requirement' of MIC at the Bhopal plant. This was confirmed by S. Kumaraswami, UCIL manager in New Delhi (Everest, 1985, p. 32). The extra MIC was sold by UCIL to other chemical firms in India. Again Bhopal mirrored Institute, where UCC held large stocks of MIC, which it supplied to other pesticide makers such as FMC and Du Pont. The other major reason for MIC storage was to ensure continued production: storage of MIC prevents any breakdown in MIC production from halting production of the finished pesticide.

The second blow to transfer blaming came on 25 January 1985, when US Congressman Waxman published an internal UCC report dated September 1984. This report warned that 'a runaway reaction could occur in the MIC unit storage tanks' (at Institute), and that 'the planned response would not be timely or effective enough to prevent catastrophic failure of the tanks'. While several instances of water contamination of the Institute tanks were handled without problems, 'they may have created a degree of confidence or lack of concern that could allow a situation to proceed to the point where it is not controllable'. As MIC tanks at Institute had been contaminated with water and other reactants several times, the report called for 'all operating personnel' to be warned. It noted: 'The fact that past instances of water contamination may be warnings rather than examples of successfully dealing with problems should be emphasized to all operating personnel.'

While UCC took action on this report at Institute, by increasing sampling of tanks for possible contamination and changing training practices, it did not notify the Bhopal plant of what it described as a 'hypothetical scenario'. Explaining why the report hadn't been passed on, Jackson Browning, UCC vice-president for safety, health and environmental affairs, said the Institute plant was completely safe and that there was no need to warn the Indians. UCC officials pointed out that the cooling system in the tanks at Institute was different from the cooling system in Bhopal's MIC storage: thus the hypothetical scenario did not arise in Bhopal. This implied that UCC had a greater knowledge of the Bhopal plant three months prior to the killing than it admitted having after the killing, when it claimed it did not know how Bhopal engineers had implemented the design standards UCC had supplied.

Nevertheless, UCC stuck to blaming the Indians. At the launching of UCC's March 1985 report on the accident, UCC chief executive officer Warren Anderson insisted that he did not know why safety devices were not working: 'Non-compliance with safety procedures is a local issue', he claimed. Anderson alleged the Indian managers were solely to blame for the disaster: 'It's their company, their plant, their people.' Thus UCC maintained that plant safety was primarily the responsibility of local plant management. While Anderson said UCC and UCIL 'are both in this together', they weren't in it together to the extent of UCC accepting legal liability for the killing. Anderson stated, 'No one at UCC knew that the Bhopal plant was being run when it was completely unsafe.'

He also claimed changes that were unauthorized by the parent company had been undertaken at Bhopal: 'The [investigation] team was shocked at what they saw. Equipment was not running. It was a surprise to the engineers who worked on the start-up.' The UCC team found that the safety system had eight serious malfunctions and admitted that just about everything that could go wrong did go wrong. But UCC 'can't be there, day-in, day-out. You have to rely on the people you have in place. My board doesn't know what valve is on or off in Texas City.' Anderson's final message, along with the ritual expressions of his own personal distress at the tragedy, was: 'Safety is the responsibility of the people who operate our plants. You can't run a $10 billion corporation all out of Danbury.'[11]

Who's in Control?

Capital found UCC's arguments useful, as its apologists eagerly used the Bhopal killing as more ammunition in its battle against state restrictions on multinational capital. For example, New York University's Gladwin and Walter suggested that 'regulations driven by nationalism may lie at the core of any in-depth explanation of Bhopal's tragedy' (*Wall Street Journal Europe*, 21 January 1985, p. 6). Having swallowed UCC's position hook, line and sinker, they wrote that

a tentative reading of the publicly available evidence so far suggests that the Bhopal facility may have been operating quite independently of the parent with regard to industrial and environmental questions – thus adding credence to the recent admission by Warren M. Anderson, Carbide's chairman, that Union Carbide India Ltd basically 'operated as a separate company'. And much of this relative autonomy can probably be traced to the pattern of restrictive Indian regulations imposed on foreign investment and the importation of products, know-how and managerial and technical skills.

This 'explanation' of Bhopal is summed up in the conclusion by *Chemical and Engineering News* (8 April 1985, p. 10): 'It is the bitter fruit of indigenization, the consensus now reads, that was basically to blame for the Bhopal disaster.'

This consensus existed only in the boardrooms of multinational capital and in its various publicity organs. This industry position is well put by an unidentified Du Pont executive:

[Developing] countries have a long way to go and they pose great technological risk to us. What has to be done is that when we build a plant in a developing country we have to install equipment that is going to be failsafe for at least ten years. And there is no way that we can do that. So without managerial control we don't go in. (Chemical and Engineering News, *8 April 1985, p. 10, emphasis added)*

Note how this executive succeeds in standing the world on its head. Instead of toxic technology posing a great risk to peripheral nations, peripheral nations supposedly pose a threat to the owners and originators of toxic technology in the metropolitan countries.

This racist response was coming from capital as early as two weeks after the killing. The *New York Times* reported:

The question of safety at overseas facilities is a particularly thorny one. Many Third World countries, including India, Brazil, Chile and South Korea, do not allow foreign companies to maintain full ownership of their plants. Many impose severe restrictions on importing equipment. And they insist that the plant hire only local people, even if they are not as well trained as their American counterparts. And, perhaps more troublesome, local management often does not have what Dr Utidjian of American Cyanimid calls the 'North American philosophy of the importance of human life'. (16 December 1984, section 3, p. 30).

This argument typically ignored many basic facts. While Bhopal came under India's Foreign Exchange Regulations Act, which limits foreign investors to a 40 percent equity stake, UCC had persuaded the Indian government in the late 1970s to grant them an exemption based on 'significant export volume and the technological sophistication of its operations'. Indeed, UCIL was one of the few firms in India in which the parent company was allowed to maintain a majority interest. Similarly, the inference that substandard Indian sources for materials had to be used is belied by the fact that in 1982 over 30 percent of UCIL's raw materials, spare parts and components were imported: thus the import conditions could not be described as stringent. Training for Bhopal management had also been provided in the USA: 'Tear-gas' Mukund, works manager at Bhopal, for example, had previously worked at UCC's plant at Institute, West Virginia.

Despite these inconvenient facts, various chemical multinationals reacted in the same vein, saying that in future they would demand much greater control of joint ventures. According to Gladwin and Walter, 'the message is that if developing countries continue to insist on a dilution of multinational corporate control, they will also be diminishing the motivation and capacity of companies to invest and to transfer environmental management and safety competence' (*Wall Street Journal Europe*, 21 January 1985, p. 6). Thus UCC was portrayed as a shackled angel, whose motivation to transfer its undoubted environmental and safety competence was diminished by a nasty, nationalist Indian government.

The best way to demolish this image is to examine UCC's operations in another peripheral country where no such government regulations impeded UCC's full control of its subsidiary. We are lucky to have such information from a US journalist on UCC's Cimanggis battery plant near Jakarta in Indonesia (Wyrick, 1981). Indonesia is a strong authoritarian state, with some 100,000 political prisoners, which does all in its power to help multinational corporations. Strikes without government permission are forbidden and only one national union is legally sanctioned. Safety and environmental laws are rarely enforced: Indonesia has 300 labour-law inspectors for 110,000 companies. Wyrick quotes a government 'invitation to investors' which emphasizes, 'One of our greatest assets is our industrious and willing people combined with your guaranteed FREEDOM TO MANAGE' (emphasis in original). Here we have a situation where multinational corporations aren't encumbered by government regulations that tie their hands and lessen their interests in their subsidiaries' operations, thus leaving them free to lavish their normal well-known care and attention on worker health and safety matters, environmental and general safety problems.

Yet UCC's operations in Cimanggis could hardly be described as a model of multinational cleanliness. The local company's own health officer claimed that at one stage 402 workers out of 750 at the plant were suffering from kidney disease related to work-place exposure. Well water at the plant was reported to contain 547 ppm of mercury. The US Environmental Protection Agency's 'acceptable' figure for US water supplies is 2 ppm. Nor did the company show that well-known corporate desire to inform its workers of health hazards at work. The company health officer was forbidden to tell the workers the water was contaminated. The company's health officer resigned two years after beginning to work for UCC, saying the company lacked professionalism.

UCC's Indonesian subsidiary showed a similar concern for secrecy, as did UCC's Indian subsidiary. Corporate officials in Indonesia refused to be interviewed by Wyrick about the plant, except to say, 'By Indonesian standards, we operate a very clean company.' They also refused Wyrick permission to inspect and photograph the plant. They were backed up by UCC management in the USA, who refused Wyrick access to UCC battery plants in the USA and who said that 'answers to questions concerning workers' safety could not or would not be given'.

The similarities between restrained Indian and unrestrained Indonesian operations do not stop there. Assistant personnel manager for UCC in Indonesia was Ashmy Hasan, whose brother was secretary of the government-approved All-Indonesian Labour Federation, as well as being a member of the Indonesian parliament and leader of its labour commission. In 1980, sixty UCC workers appeared before that parliamentary labour commission to complain about working conditions and lack of attention to safety at the Cimanngis plant. This appearance resulted not in reforms at UCC but in all sixty workers losing their jobs without explanation.[12] UCC's closeness to the government also allowed it to avoid the legal ceiling of fourteen hours overtime per week; it received special permission for a seventy-two-hour week. According to the subsidiary's health officer, workers who refused overtime 'were advised by Union Carbide to resign' (Wyrick, 1981, p. 24R).

UCC's unrestrained operation may be best summed up in Wyrick's account of the death of Haryanto, an apprentice worker at the plant:

The examining doctor reported that the body was black with carbon dust.

Haryanto worked at the machine that mixes carbon black and other chemicals to form the inside body of an Ever Ready battery, a battery like the ones in your flashlight or transistor radio. Because the two dust-collecting machines in the 'mixing room' were broken and had not been repaired – just as the leaky drain had not been fixed – Haryanto worked in a haze of black dust.

A probationary employee on the job only two months, he had been required to work three consecutive overtime days and was extremely fatigued at the time of his death, 'too tired to think properly', Union Carbide's medical director for the plant said.

Haryanto was unable to ask for guidance when the mixing machine gave him trouble. 'There was no supervision when the accident occurred,' a company investigation reported. 'The supervisor in charge of the area was having meals.'

An electrical switch in the mixing room was 'unsafe' because it was improperly located and was unshielded, according to the company investigation report, marked 'Business Confidential'. When the mixing machine was not loaded correctly or was malfunctioning, the switch would automatically shut it down. According to the company report, Haryanto, trying to restart the mixer, depressed the cut-off switch with a naked metal saw blade. The blade 'became energized to 222 volts' as Haryanto stood on the wet floor, the report said, and he was electrocuted. (Wyrick, p. 23R)[13]

A further point must be made regarding the transfer of safety technology. UCC's licence to operate the Bhopal plant had been recently renewed on the basis that it would be provided with the most up-to-date safety technology. According to the *Indian Express* (7 December 1984), 'In its application in September 1982, Union Carbide India Ltd sought extension on the grounds that its parent company, having "experience in handling toxic chemicals over several years", would make available to UCIL "the current knowledge and experiences in handling highly toxic materials" on a continuous basis. It said that "continuous availability of data in this area will assist Union Carbide India Ltd in fully protecting the plant personnel and properties".' UCIL further justified the extension on grounds that as a result of the collaboration, the US firm would make available 'toxicology data on products produced besides antidotes and safety precautions'.

Similarly the Indian case against UCC submitted to Judge Keenan on 8 April 1985 charged: 'Defendant Union Carbide represented to the plaintiff [i.e. the Indian government] that it would provide the Bhopal plant with the best and the most up-to-date technical data and information in its possession for the manufacturing, processing, handling and storage of MIC and that it would continually update this information' (APPEN, 1985, p. 225).

CRACKS IN THE DEFENCE

To accept UCC's claim that it was not in control of UCIL requires a largish suspension of disbelief. UCC had been permitted to maintain a controlling interest in UCIL owing to the allegedly high-tech nature of the industry.

Why did UCC insist on a controlling interest if not for purposes of control? Veerenda Patil, the Indian government's chemicals and fertilizers minister, in response to UCC's transfer blaming, told the Indian Parliament on 29 March 1985 that the government had adequate evidence to establish the company's culpability. Dr S. Varadarajan, the Indian government's chief scientist, said his staff had been told that, after discussion with the US headquarters of UCC, Bhopal managers had concluded the refrigeration unit to chill the MIC was unnecessary. Dr Lal, chief medical officer at Bhopal, observed: 'The safety precautions we took were the best possible. *We did everything the Americans advised*. In fact we used to think that we were overdoing the safety.'

Some indications of UCIL's concept of safety may be obtained from its personnel manager, B. R. D. Krishnamoorthy, who proudly told Agnew (reporter for *Safety and Risk Management*) of UCIL's continuous concern for safety:

You know we had a meeting of all managers in the plant every morning, about a dozen of us and most of the time we discussed safety. I give you an example. Each workman had his own locker, and one day somebody straightened up and banged his head on the open door. We discussed for a long time how that could be avoided. Another time when someone slipped in the showers we investigated how to prevent a reoccurrence.

Another example of UCIL management's devotion to safety can be found in the fact that they advised Bhopal workers to 'develop' resistance against toxic substances by drinking six or seven glasses of milk a day and eating a high-protein diet of fish and eggs (*Economic & Political Weekly*, 15 December 1984).[14]

The strategy of transfer blaming received another set-back in October 1985. UCIL documents, filed by Indian government lawyers in the USA in a motion for further discovery of UCC documents, showed UCC had been considering closing the Bhopal plant and shifting the operation to either Brazil or Indonesia months before the killing. UCIL had been asked to undertake a preliminary study of the costs of dismantling the MIC and other pesticide production units at Bhopal. Interestingly, UCIL was studying closing the plant at the time when key safety devices were turned off or left unrepaired. Indeed, the question of whether to sell the Bhopal plant reached Anderson's desk, where the sell move was approved shortly before the killing. As the *Wall Street Journal Europe* commented, this 'indicates that the unit [UCIL] was less than a distant cousin'. UCC denied it intended to sell the plant: 'The company denies it wanted to shut it, saying rather that it intended gradually to switch away from the unsuccessful carbaryl insecticides, which it sold under the Sevin brand name, to other products based on the same technology' (*Financial Times*, 7 December 1984).

UCC documents submitted for examination under the US legal process of document discovery, whereby documents in the company's possession relating to the issue under trial were made available to the plaintiffs, showed further cracks in UCC's defence. Stanley Diamond reported in the *New York Times* that these documents, many stamped 'confidential', appeared to contradict UCC's claims of non-involvement in the Bhopal project:

After a 1981 accident in Bhopal that killed a worker, a telex said that improvements 'will receive close attention by the management committee in New York' and that it was 'very essential' this committee know the 'specific actions' to prevent recurrence. Another memo said, 'No design changes have been made without the concurrence of general engineering or Institute plant engineering', referring to Carbide's corporate engineers in Institute, West Virginia. (3 January 1986, pp. D1, D3).

UCC also claimed in court papers that the plant design that was originally supplied by an Institute team was changed in India. 'However, Warren J. Woomer, a Carbide engineer, said in an affadavit that he had approved the design by tracing "every line, every valve, every instrument" when the plant started up' (*New York Times*, 5 January 1986, p. D3). Contrary to other UCC claims that changes in design at Bhopal since the start-up were unauthorized, it was reported that the Indian Central Bureau of Investigation inquiry had found that UCC had approved in May 1984 'a design change in the pipelines leading to the MIC storage tanks which was directly responsible for the water entering tank number 610 on the fatal night' (*Sunday*, 7–13 April 1985, p. 19; also Agnew, 1985, p. 17).

UCC is also reported to have done feasibility studies on whether to build the plant at Bhopal. It has been reported that, while the studies were being performed, concern was expressed over whether proper maintenance could be maintained in India for such a complex plant. Roger Aitala, a former UCC chemical engineer who worked on one of the studies, told *Business Week* (10 January 1986, p. 23), 'There is no question that they [UCC] knew what they were dealing with.'

Nevertheless, UCC continued to maintain that UCIL was 'a "free-standing entity" with nothing but a voluntary association with Union Carbide' (anon., *Business India*, 1986, p. 77). In depositions taken for the plaintiffs in the US court case, UCC's management denied central control by describing their own manuals as 'gobbledy gook' and by offering such contorted arguments as the following:

Question: *In order to secure effective management control of an affiliate Union Carbide need not have 100 percent say on the board of directors, correct?*

Answer *(James M. Rehfield, UCC executive vice-president and UCIL director): Union Carbide does not control its affiliate companies, period.*

Question: *Sir, who controls an affiliate company?*

Answer: *The board of that company.*

Question: *Who is the board elected by?*

Answer: *The equity participants.*

Question: *And who's the majority equity participant in Union Carbide India Ltd?*

Answer: *Carbide's 50.9 [percent]. (anon, Business India, 1986, p. 77)*[15]

In court cases against UCIL in Bhopal, its own lawyers have stuck rigidly to the UCC line that the American company does not control the Indian company. UCIL's chief outside counsel, Vijay Kumar Gupta, told the *American Lawyer:*

Union Carbide India Ltd is an Indian company regulated under the Indian Companies Act. Therefore the Union Carbide Corporation cannot be held liable for any of the acts of the Indian company. It is again wrong to say that the American company is the parent company. It is not the parent company. It holds some shares. It is not the proprietor . . . The Indian company has nothing to do with the US company. (April 1985, p. 130)

When the Indian company is following so faithfully a line devised by top corporate management to protect UCC, it is obvious that UCC's writ still runs at UCIL.

Sabotage Theory

If UCC's first line of defence seemed untenable, its next line of defence approached the ludicrous. Even the business press exhibited a slight disdain for what it correctly designated UCC's 'sabotage theory'. Speaking on the subject at the March 1985 UCC press conference, chief executive officer Warren Anderson 'offered a scenario in which a disgruntled worker might have wanted to sabotage the plant without actually meaning to cause the ensuing tragedy'.

Extrapolating on only a slight hint in the technical report that sabotage might have been the cause, Anderson said the leak 'could have been inadvertent or it could have been deliberate'. To back up its sabotage theory, despite the lack of any evidence, UCC officials said it was difficult to conceive that up to 240 gallons of water (which they estimated was needed to trigger the runaway reaction) could have entered the tightly sealed tank totally by accident. Another UCC spokesperson said the same day, 'We have never had a problem there before. There are no radicals or groups like that. I can't think of a motive.' Indeed, some US financial analysts dismissed UCC's theory as nothing but, in the words of K. S Raman, 'a carefully orchestrated attempt to influence the upcoming legal hearings'. Anderson himself admitted to a US Congressional panel that he had 'no evidence whatsoever that sabotage was behind' the disaster.

Despite the lack of evidence and any motive, UCC clung tenaciously to its sabotage theory, no doubt influenced by the fact that sabotage was the only argument UCC could use to minimize its own liability. In a statement to investors in June 1985, it said 'UCC does not believe that it will be or should be liable for the disastrous events at Bhopal. We were never able to rule out sabotage and still have no information that suggests the direct cause was other than sabotage. The Bhopal plant, as designed, met all appropriate safety standards.'

Similarly, at the 'Chemical Industry After Bhopal' conference held in London in November 1985, UCC's vice-president for health, safety and environmental affairs, Jackson Browning, was reported by the *Guardian* (8 November 1985, p. 11) as saying: 'The company's scientists had established *beyond doubt* . . . that the introduction of 120 to 240 gallons of water into a storage tank could not have occurred by accident' (emphasis added). He further contended UCC had 'all but ruled out anything but a deliberate act'. Finally the nefarious and nebulous saboteurs became a shadowy Sikh terrorist group, the Black June Movement, unknown until UCC first reported their existence. This theory became even less credible a defence when its source became known: some people in one Punjab city were reported to have seen a poster from this Black June Movement claiming responsibility for the disaster. Shri Ahmad, India's commercial consul in New York, commented acidly: 'On the basis of a single poster, seen by only a few people in one small city, Union Carbide, one of the largest companies in the world, has built up an entire theory to defend itself' (*Chemical Week*, 18 December 1985, p. 13).

What is particularly worth noting is that UCC had now narrowed the definition of the accident to one simple action – the entry of water into the storage tank that contained MIC. This is a typical management tactic[16] in dealing with chemical accidents, described by one industry consultant (Howard, 1983) as not looking for the *truly basic* cause. In this case, UCC was stopping its analysis of the fault tree at the sapling stage. At the November 'Chemical Industry after Bhopal' conference, UCC also dismissed the water-washing theory, advanced by the workers and by the international trade union report on Bhopal, of how the water had entered

UNION CARBIDE'S RESPONSE   47

Here is what investigators have learned:

A few hours after the incident, a hose with water running out of it was found lying near the storage tank.

In addition, several witnesses discovered that a pressure gauge had been removed, leaving an unplugged opening into the tank.

Additional evidence supports the employee sabotage finding, including: a shift instrument log indicating pressure gauges were present a few days before the incident, a sketch reflecting that water had entered the tank through a connection to the pressure gauge, logs that were altered or missing, and statements from other witnesses about the incident.

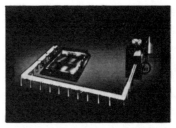

Previously, some people thought that routine washing of a filter system far from the MIC storage area caused water to back up through 400 feet of pipe and enter the storage tank.

And finally, at the Indian Government's direction, a hole was drilled into the pipeline's lowest point. Had water gone into the pipe, it would have remained because there was no way for it to evaporate or escape during or after the incident.

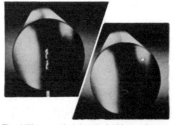

The drilling revealed that the inside was bone dry.

The evidence is overwhelming that the tragedy at Bhopal was an act of employee sabotage.

Promoting its sabotage theory, UCC's glossy, full-colour brochure reduces the entire issue to how water entered the MIC storage tank.

tank 610: 'Our team rejected that theory in March as being highly improbable and nothing has been uncovered since that has changed our mind.' Indeed, Jackson Browning told the November conference that UCC scientists had established *beyond doubt* that the water could not have been introduced into the tank by accident.

Compare this position with what Ron Van Mynen, corporate director of safety and health and chairperson of the UCC Bhopal investigation team, actually said in March 1985:

We think it's more likely *that the water travelled from the nearby utility station directly into the tank. The nitrogen and water lines are at the same utility station.* If someone had *connected a tubing to the water line instead of the nitrogen line either deliberately or intending to introduce nitrogen into the tank, this* could *account for the presence of water in the tank. (Van Mynen, p. 12, emphasis added)*

UCC admitted in March 1985 that its investigations of the killing were severely limited by the fact that they were unable to interview the workers who were on duty the night it happened. At the 'Chemical Industry After Bhopal' conference, Browning admitted that UCC's investigatory staff were forbidden admittance at all for four days and only thereafter were allowed to the storage tank. All in all, the team had around a week to investigate. The Indian journal *Sunday* reported (7–13 April 1985) that the UCC investigating team was not allowed to interview any of the Bhopal workers, nor was it allowed access to plant documents. The team is reported to have met only two persons – the plant's production manager and works manager – who could be of any assistance. Neither of these persons had been at the plant on the night of the killing. Both arrived after the leak had taken place.

More interesting yet, UCIL managing director Gokhale admitted that UCC had not been able to examine the pipelines to determine how water got into the tank (Everest, 1985, p. 187). Therefore UCC's March report had been based on analysis of the chemical residue left in the tank after the runaway reaction and the leak. From this residue, UCC attempted to reconstruct the chemical reactions that could have led to the residue. The industry press noted that UCC's March 1985 report left many questions unanswered, while US environmentalists condemned it as highly speculative. Since that report, UCC had performed further analysis on the residue from tank 610 – analysis which it claimed confirmed its previous findings. From March to November, UCC's only investigative work consisted of chemical investigations; on what basis did they decide the water could have entered the tank only from the nearby utility station? The workers had said no connection had been made that night from the utility station to tank 610.

UCC's dismissal of eyewitness accounts in favour of a theoretical reconstruction is a normal management tactic[17] – after all, they imply, they are only Indian workers and cannot be trusted. This left UCC with its scientific analysis of the residue of the chemicals in the tank that leaked as its only evidence. On the basis of that evidence, how did UCC move from what its scientists had previously presented as possibilities or probabilities to such a definite certainty? Hardly on purely scientific grounds.

UCC was prepared to dismiss the workers' and trade-union account of the direct cause in March 1985, saying, 'However, entry of water into tank 610 from this washing in the MIC unit would have required simultaneous leaks through several reportedly closed valves, which is highly improbable.' Yet it also admitted that at Bhopal everything that could go wrong did go wrong. UCC described the mass murder thus: 'The incident [sic] was the result of a unique combination of unusual events.' Given this, why was the workers' explanation any more 'highly improbable' than UCC's explanation? The workers' explanation was backed up by their evidence that valves continually leaked and also by UCC's own admission that maintenance at the plant was appalling. UCC presented no evidence for the hidden hand it had introduced as the causal factor. It is obvious that UCC's own 'highly improbable' account of the cause of the leak stems from UCC's interests in denying liability. The claims that their scientists had dismissed other accounts of the leak on a scientific basis was nothing but an attempt to clothe UCC's economic interest with scientific legitimacy.[18]

Anyone who reads UCC's March report on the disaster (UCC, 1985b) will find more than enough science in it to legitimate anything. The scientific content of the report – intimidating as it is to anyone but chemical engineers – is, of course, only one side of the story. The report is a theoretical reconstruction of what UCC's scientists believe happened during the runaway reaction inside the MIC storage tank. Thus an alternative explanation of the accident in chemical terms is possible.

This is most obvious in the still highly controversial question about the reaction: whether MIC had decomposed into hydrogen cyanide. Following the leak in Bhopal, autopsies performed by Dr Heeresh Chandra at Hamidia Hospital found symptoms of cyanide poisoning. (A major controversy arose over this issue, which is covered in Chapter 3.) In response to claims that hydrogen cyanide was among the gases that leaked from the Bhopal plant, a UCC spokesperson told the *Observer* (10 November 1985), 'Based on our research and observations, we *feel* you have to get MIC up to extremely high temperatures – in excess of 350 degrees centigrade – before you get any appreciable amount of hydrogen cyanide. Our technical people *felt* that the temperatures in tank 610 *probably* reached between the 200 and 250 degree range' (emphasis added). Later another UCC spokesperson told the *New York Times* (1 December 1985), 'The formation of hydrogen cyanide during the incident is not possible.'

Against this, N. Karin Ahmad of the US National Resources Defense Council contended that his studies indicated the runaway reaction could have reached 840°F (449°C), the temperature at which hydrogen cyanide is created in his opinion. Anand Grover, a member of the Bombay Lawyers' Collective, a legal group involved in litigation over Bhopal in India and himself a chemical engineer, said the temperature reached during the runaway reaction could have been as high as 1,000°F (538°C). The official Indian government report (Varadarajan *et al.*, 1985, p. 11) reads: 'From the products found in the residue, the calculated amount of heats of chemical reactions and the extent of bulging of the exhumed tank, it is surmised that the temperature in the tank rose above 250 degrees centigrade at the time of the accident.' Elsewhere UCC stated that at temperatures of about 400°C, MIC decomposes into nitrous oxides, carbon monoxide and dioxide – and hydrogen cyanide. To further complicate matters the *Madhya Pradesh Chronicle* (26 March 1985, reprinted in Pinto and O'Leary, 1985, p. 65) quotes confidential UCC manuals as saying thermal decomposition of MIC takes place at 218°C. UCC says the reaction did not rise above 200°C: the intactness of tank 610 proves this. Indian scientists responded by saying the complexities of such a runaway reaction could create 'hot spots' with much higher temperatures. UCC, naturally, rejected this.

Similar controversy exists over the volume of water that triggered the reaction. UCC said 120 to 240 gallons were required. In January 1985 the Indian government's chief scientist Varadarajan, head of the CSIR, was quoted by the Press Trust of India as saying 'just half a kilogramme (about 1.1 pints) of water' triggered the reaction (*Guardian*, 5 January 1985, p. 4). His December 1985 report said up to 110 gallons could have been involved.

The final controversial point regarding UCC's report on the accident relates to the composition of the released toxic gases. UCC's March 1985 report makes no attempt to analyse what was the composition of the gases that finally leaked. The report blandly states (UCC, 1985b, p. 24), 'Approximately 54,000 lb of unreacted MIC left Tank 610 together with approximately 26,000 lb of reaction products.' Indeed, in its desire to ignore the escaped reaction products, the report's summary (UCC, 1985b, p. i) omits them entirely. Despite the hundreds of reactions UCC performed to come up with its description of the reaction, none of the gaseous products of the reactions are analysed. Everest reports (1985, p. 177) that UCC did not respond to questions on its knowledge of what the escaped reaction products were. Particularly enlightening here is UCC's account of what products escaped after the runaway reaction in Institute in August 1985: the detailed list (see *Chemical and Engineering News*, 2 September 1985, p. 6) contains twenty-three compounds, in amounts ranging from 650 lb down to 7 lb. The major difference in Institute was that a US environmental agency performed an analysis of the leak's constituents and published it: UCC was therefore forced to provide more information on the constituents of the toxic gas cloud. Needless to say, the analyses by UCC and the environmental agency disagreed.

Some work on the issue was undertaken by Indian state agencies. Dr Misra of Gandhi Medical College in Bhopal told Everest (1985, p. 73) that the Indian Council on Medical Research had proved conclusively three types of gases besides MIC were liberated during the leak: hydrocyanic acid (also known as hydrogen cyanide), nitrous oxide and carbon monoxide. Tests undertaken by Dr P. K. Remachandran and reported at an Indian Council on Medical Research meeting on 4 April 1985 found that at 200°C, the decomposition products of MIC contained 3 percent hydrogen cyanide, while at 400°C the proportion of hydrogen cyanide was 20 percent (APPEN, 1985, p. 137). In June 1986, UCC's medical director,

Dr Bipin Avashia, admitted that there could have been hydrogen cyanide in the toxic cloud that enveloped Bhopal (*New Scientist*, 28 November 1985, p. 41).

Regarding the sabotage theory, and the shadowy alleged Sikh terrorists who were too modest to publicize their success widely, they must have been highly knowledgeable about chemical engineering. Varadarajan's report to the Indian government in December 1985 notes: 'The scientific analysis shows that any addition of water alone, even deliberately, could not lead to such an accident. Anyone wishing to cause an accident of this nature would have to be presumed to have very substantial knowledge and information that metal contaminants would already be present and that the alarms and safety systems installed for containment were grossly inadequate.' Given the sophisticated chemical-engineering knowledge UCC has ascribed to these terrorists, one wonders why it does not solicit advice from them.

Return of the Disgruntled Worker

UCC's sabotage theory later abandoned those shadowy Sikhs with chemical engineering degrees, only to return yet again to the 'disgruntled worker' theory. Hilfra Tandy, a reporter with the Financial Times publication *World Petrochemical Analysis*, was on an all-expenses-paid trip to the USA to view UCC's plants there. While in UCC headquarters in Danbury, Connecticut, she was leaked the latest version of the sabotage saga. Harvey I. Colbert, UCC's media-relations manager, told the *Sunday Times* (10 August 1986):

Our investigations to date demonstrate that the Bhopal tragedy was a deliberate act.

These investigations are now focusing on a specific individual employee who was disgruntled and who had ample opportunity to deliberately inject the large amount of water into the storage tank which caused the massive leakage.

UCC's theory now is that the (unidentified) disgruntled worker rigged up a water hose to the storage tank, intending only to spoil a batch of the chemical, following a row with his supervisor. Another UCC spokesperson told the *Guardian* (11 August 1986, p. 7) that no political motive was involved, but refused to give further details, as the matter was before the courts.

While UCC was drawing attention to this suspect worker, it was as usual attempting to have it both ways. Harvey Colbert told the *New York Times* (11 August 1986, p. A4) that UCC had 'never used the word sabotage and still is not charging sabotage'.[19] Sources 'close to the company' filled out the picture for the *New York Times:*

> ... the man was an employee who had been demoted a week earlier and was at the plant on the night of the accident without management authorization. They added that some documents had been altered to change key details that would determine culpability, and that the company was closely questioning the employee and others who worked there when the leak occurred. The plant is now closed.
>
> One source said the information was not provided to the Indian government because the company had not yet 'closed the loop' in its investigation: the employee had not confessed and witnesses would not pin the blame on the employee. *(12 August 1986, p. A3)*

UCC refused to tell the *Chemical Marketing Reporter* (18 August 1986, p. 3) 'how long the individual has been a suspect, or whether he was one of the eyewitnesses Carbide investigators interviewed immediately after the gas leak'.[20] UCC's counsel Bud Holman suggested the Indian government withhold its own evidence of sabotage as it would hurt India's court case against UCC.

Reaction varied, but was mainly disbelief. G. G. Nayak of the Indian Chemical Manufacturers' Association told *European Chemical News* (25 August 1986, p. 4), 'We do not believe it was sabotage.' A US lawyer for the Indian government also reacted with scepticism: 'If they are really serious about their story, they should prove it instead of making vague statements, and they should give the name of their suspect to the proper Indian authorities. There is no factual basis to support this story and, let's face it, Union Carbide is not a company with a long record of credibility' (*Guardian*, 12 August 1986, p. 6). Here, we have UCC not as the little boy who cried wolf, but as the little boy who changed his story about which wolf it was – too often to be believed.

2 The Indian State's Role

It was unfortunate for the people of Bhopal that the killing occurred during a general election campaign, as it was used for immediate political purposes during the election. 'With elections around the corner, candidates and political parties mobilized their workers for relief work, publicizing their contribution in the process' (*Madhya Pradesh Chronicle*, 5 December 1984). Bhopal victims were mobilized to shout slogans praising Gandhi on his electioneering visit to the city, despite the fact they were acutely ill. Both parties accused each other of responsibility.

In the longer term, Indian politicians had more important things to concern themselves with than the people of Bhopal. Even the conservative paper the *Hindu* was forced to comment: 'Among their preoccupation with the election campaigns, the Central State ministers had no time to provide the kind of higher direction required in organizing relief on a large scale, besides investigating the cause of the mishap [sic] and taking steps to avoid the recurrence of such industrial accidents.' Thus the citizens of Bhopal were denied even the dubious benefits of organized government's reputed ability to react effectively in relieving a disaster.

If the state was too busy to provide adequate relief for the killing's victims, it worked diligently to help cover up the extent of the killing.

THE INDIAN STATE'S ROLE 55

After the Bhopal killing, the government offers platitudes instead of financial aid: 'Despite all our differences of religion, differences of language and other differences, we are all children of India, and if we go forward together, then India can become strong' – Indira Gandhi; 'We must give the highest priority to the rehabilitation of the gas victims' – Rajiv Gandhi.

Official figures for those murdered by the gas leak hover around 2,000 but unofficial sources claim 8,000–10,000 died. By 5 December, according to state officials, 1,267 people had been cremated or buried. By 15 December the official death toll stood at 2,500, but residents of Jaiprakash Nagar claimed that 3,000 were killed in that area alone. Similar claims that a cover-up operation had been mounted were made regarding the Pemex gas explosion on 19 November 1984 at Ixhuatapec, Mexico City: official figures say 400 were killed, but local residents said bulldozers buried hundreds more victims.[1]

One doctor at Bhopal's Hamidia Hospital said: 'As soon as a patient was declared dead, his relatives would just vanish with the body: I saw at least fifty bodies taken away like that. I would estimate that anything between five hundred and a thousand bodies were taken away before their deaths could be registered.' The Catholic relief agency Caritas said 20,000 people died. United Nations Children's Fund staff estimated more than 10,000 people died. The India Cloth Merchants' Association said it distributed 10,000 shrouds for Hindu and Muslim death services. V. Sriram, an official with the Federation of India Chambers of Commerce and Industry, said, 'So many families were wiped out that it isn't difficult to imagine a number as high as 20,000' (*Chemical and Engineering News*, 21 January 1985, p. 4).

In *Health & Safety at Work*, the Union Research Group (URG) describes the state's immediate response:

Concealing the death toll involved not only censorship but actual destruction of evidence: thousands of bodies were cleared away and cremated or dumped in the Narmada river . . . it was not even ascertained that the people were dead; those who regained consciousness after being thrown in the river came back to relate their experience. But we will never know how many were burned alive.

Other cover-up activities included

preventing voluntary organizations from entering government relief camps, closing down government relief camps while there was still an urgent need for them; discharging patients from hospitals while they were still seriously sick; withholding information obtained from autopsies on the real effects of the gas; changing the 'cause of illness' on hospital case papers from 'MIC poisoning' to something else, so that compensation could not be claimed. (Health & Safety at Work, *May 1985, p. 41*)

Bhopal presented a major change in strategy in dealing with toxic disasters, in that no evacuation to escape contamination was ordered. The Press Trust of India quoted scientists at Lucknow Industrial Toxicology Research Centre as calling for affected areas to be evacuated, since the gas was likely to remain in the atmosphere for three or four weeks. B. Pathak, Professor of Chemistry at Calcutta University, believed that the atmosphere in the affected areas should be sprayed with diluted ammonia. But it was obvious that the Indian government had no intention of evacuating the city, as the costs of such an evacuation would have been immense.

In previous toxic catastrophes, a major controversy had always arisen over the physical boundaries of the area considered to be so heavily polluted, or in such immediate danger of pollution, that it needed to be evacuated. In Seveso, there was an official division of affected areas into Zones A, B and C. At Love Canal, where a neighbourhood had been built on a toxic waste dump, the state drew a line beyond which it would not purchase residents' houses. At Three Mile Island, the state vacillated over who was to be evacuated and from where. These were major issues, as well as ways of dividing and reassuring the affected populations. (See Levidow, 1979, and Pomata, 1979.)

When no real evacuation was likely in Bhopal, the Indian state's desire to limit the disruption showed in its immediate denial that pollution existed to an extent that would warrant evacuation. Thus, eight hours after the leak, the atmosphere in Bhopal was declared free of the gas, though people were warned they should be careful about what they ate for another seventy-two hours (*Financial Times*, 6 December 1984). On his electioneering visit to Bhopal, Prime Minister Rajiv Gandhi said the water had been tested and no toxic substances found in it, though the government's chief scientist Varadarajan said later that testing only began on 5 December. Gandhi also 'said emergency steps had been taken to stop the spread of toxic effects of the gas on animals and crops. He gave no details' (*New York Times*, 5 December 1984, p. A1).

Praful Bidwai drew a revealing parallel in his recollection of his visit to Seveso and spoke of the 'frightening optimism' of the Indian authorities:

They have already announced that everything is fine, that vegetables and water are safe. I am just looking at a sample of diseased spinach – one of the many vegetables within 2/3 km of the killer plant. I can see that portions of it are corroded and there is a white deposit on the leaves.

In Seveso, 730 people were evacuated from 'Zone A', the worst affected, although no one had died. Here no one will even consider evacuation, or a ban on sale and farming of vegetables or meat or milk. Arjun Singh says it's for the scientific experts to decide. But isn't that a political decision really? How can scientists decide what is safe and for whom? Besides, they haven't yet started to analyse samples in any serious way, they don't have the instruments to measure the tiny amounts of toxin that water, milk and

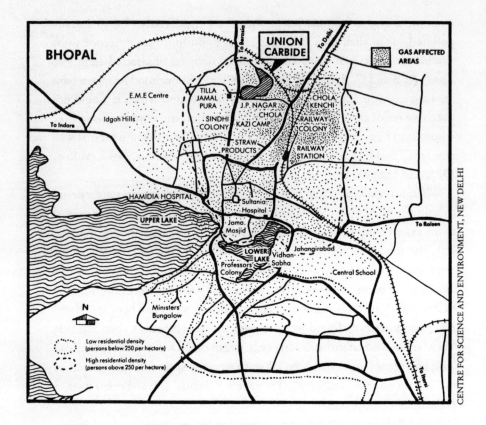

The poison gas most affected the high-density, poorer residential areas (colonies).

vegetables may contain. Strange that they should be so confident and arrogant. (Bhopal – Industrial Genocide?, *pp. 31–2*)

Within a few days of the accident, the Delhi Science Forum (1985, p. 36) reported: 'presumably on the basis of preliminary investigations, formal statements were issued that air, water, vegetables and foodstuffs were safe everywhere in the city. At the same time TV features informed people that poultry was unaffected but wanted people not to consume fish, etc. Confusion was rampant; people were asking was it safe to consume eggs, vegetables, etc.'

The Delhi Science Forum were being too charitable in suggesting these reassurances from the state were made on the basis of preliminary testing.

According to other sources, reassuring statements were issued before any testing was performed:

At a press conference on 6 December, the Chief Minister declared that the air was totally safe, but an official from the Environmental Pollution Control Board who this writer met that evening said that S. Varadarajan, head of CSIR, whose services had been requisitioned by the Madhya Pradesh government to handle the post-leakage situation, had left Bhopal that afternoon carrying air samples to be tested in either Delhi or Lucknow, since the CSIR laboratory in Bhopal lacked the necessary equipment to do the job. Tests conducted at the initiative of science students indicated the presence of MIC 'till as late as 0.15 p.m. on December 5. (Bhopal – Industrial Genocide?, p. 46)

Furthermore, while the government declared vegetables to be safe for consumption, these assurances were received by the people of Bhopal with a grain of salt:

Because it was an open secret that consignments of vegetables and fruit were arriving regularly from the government farm at Pachmari, a hill resort near Bhopal, for ministers and senior bureaucrats. In any case the pernicious effect of the gas on vegetation was manifest. Trees, plants and creepers in the vicinity of the plant were charred, and a little farther leaves had turned brown or a deep yellow. (Economic & Political Weekly, 24/29 December 1984)

Confirmation of the lack of scientific basis for scientists' assurances of safety was even provided by the scientists themselves. For example, on 5 December, Dr R. L. Rajak, adviser on plant protection (i.e. pesticides and biocides: more double-speak) to the Indian Central Agricultural Ministry, said there was no chance of toxic effects on plants, as MIC 'reacts with moisture and becomes non-toxic, yielding some by-product, besides heat', thereby supporting the UCC line. At the same conference Dr Rajak admitted no sample tests had been carried out. On 8 December, Dr Mukherjee, agricultural chemistry professor, said detailed chemical-residue analysis would be completed on the 9th. However, he did not hesitate to say vegetables were safe after washing and cooking. Results were not released on the 9th. The local Public Health Engineering Department joined in, claiming water analysis had revealed nothing alarming, but it did not present its test results.

Indian government scientists also joined the international chorus of reassurance. The *Madhya Pradesh Chronicle* (6 December 1984) quoted Dr N. A. Ayyager, senior scientist at Pune National Chemical Laboratory: 'MIC is not generally toxic to human beings except in heavy doses'; while Professor Mohan, the Indian government's ophthalmology consultant, stated (6 December 1984) that it was clear, from hospital records, that patients were in no danger of permanent blindness. A break in the chorus of reassurance came on the 11th: 'Dr R. P. Dhanda, eminent ophthalmologist and former president of the Ophthalmological Society of India, said here today that the chemical burns of the eyes, even if cured, remain active for a long time and, generally, the late effects of the burns may be worse than the initial tissue reaction' (*Free Press Journal*, 12 December 1984). But on the 13th, after visiting Bhopal with the Royal Commonwealth Society for the Blind, Dr Dhanda fell into line: 'No person affected by the gas leakage in Bhopal is likely to become blind, though some of them may have partial loss of vision . . . [after] the study of the chemical burns on the eyes, only a small number of people have been found to have suffered damage to cornea in their eyes and most of the affected were likely to be free from the symptoms with appropriate treatment.'

It is worth noting here a rhetorical question posed by a local environmental group, Eklavya. 'Why is it that many scientists refused to even consider that MIC may have serious consequences which are not even apparent? If so little is known about a chemical, is it not essential to rise to the occasion with an open inquiring mind?' The reasons are obvious. The scientists involved were prepared to make reassuring statements without studying the evidence and accumulating and analysing information. Indeed their reassurances directly contradicted the physical evidence around them, because of the political role allotted to them by the state and capital in the post-killing scenario. For Indian scientists, the fact that most of them are employed by the government and the public sector means they are wide open to state manipulation. Furthermore, most of them adhere to a pro-technology ideology.

The Centre for Science and the Environment in New Delhi has pointed out that in India 'most producers of scientific knowledge work for the government or the corporate sector, both of which close information to the public, especially in adverse circumstances'. Said D. Banerji, community health expert of the Jawaharlal Nehru University, 'The Bhopal tragedy has exposed the most deplorable state of the community of scientists in India.'

Some indications of the views of India's élite scientists can be found in Varadarajan's statements that Bhopal is the price to be paid for 'development', increases in agricultural production and 'progress' in developing countries. On the 15th he said: 'The whole issue has to be seen in the context of the cost–benefit ratio. It has to be looked at in terms of necessity. No technological operation is entirely without risks, we can't make any advances without them. We can only reduce risks.' That the chief scientist charged with handling the post-killing situation should hold these views is no major surprise. When mass murder is a 'necessity', so are censorship and lying.

For the Indian state to succeed in limiting the crisis, tight control of information was essential. This reached ludicrous extremes when information on the meteorological conditions on the morning of the killing was declared classified. To quote the Delhi Science Forum again:

The authorities have adopted a curious strategy of withholding vital information regarding the extent and the effects on human beings, birds, animals and vegetation. Much of the information generated is being treated as 'classified'. Only modified versions, often deleting all the reservations and qualifications attached to the test findings by the investigating team, are being released to the public. (1985, p. 35)

Much as UCC treated its information on MIC as trade secrets, the Indian state treated its information as state secrets. The Indian Council on Medical Research issued strict instructions to all doctors, both private doctors and government doctors, not to disclose their findings to the press or public. Post-mortem results and case histories of gas victims were declared classified. This had an adverse effect on treatment of the gas victims. For example, as Dr Bhandari of Gandhi Medical College told a member of a Japanese group investigating the killing, 'since it is prohibited by the state government to publicize the detailed data [on MIC's genetic effects], I cannot show them to you here, but we view this very seriously, and have been discussing what to do about it' (Reiko, 1985, p. 58).

The reason behind such secrecy was summed up by Arjun Singh, chief minister of Madhya Pradesh State: 'Information will spread fear. We want to arrive at complete conclusions if we are to progress in our work and treatment' (*Chemical and Engineering News*, 21 January 1985, p. 4). This is only to be expected from the minister who refused to order any action on the morning of the killing until he had the 'precise technical details'. This control also necessitated preventing or obstructing independent investigations. However, the Delhi Science Forum reports (p. 13), 'In sharp contrast to the total clamp-down on information and discouragement to Indian scientists willing to undertake independent studies, is the complete freedom with which such foreign "experts" who are known to be associated with the foreign defence research laboratories are collecting detailed information in Bhopal. They are also conducting detailed studies of their own.'

Similar controls had to be exercised on the attribution of blame for the killing. Thus the sealing of the plant by the Central Bureau of Investigation effectively sealed information about the plant and what really happened there. Reporters were denied access to the plant by the Bureau, as was the international trade-union mission in March 1985 (ICFTU, p. 7).

Finally, the immediate provision of relief was riddled by the usual corruption and bureaucracy. On his electioneering trip to Bhopal, Gandhi had promised immediate relief of 4 million rupees (£274,000) and the Madhya Pradesh state government also set up a relief fund. The immediate payments – 1,000 rupees for 'ordinary' injuries and 10,000 rupees for serious injuries after hospital evaluation – were given in the form of crossed cheques. Yet most victims had no bank accounts and were required to open accounts after depositing 20 rupees before the banks would handle their cheques. Banks had to request special permission to open accounts for those who did not have the necessary opening deposit of 20 rupees.

These payments became a vehicle for immediate political favours. Doctors at Bhopal's Hamidia Hospital went on strike after one doctor was hit by a local councillor from the Congress (I) Party, Rajiv Gandhi's party, in an argument over whether a patient should be readmitted to hospital: the councillor wished to have the patient readmitted so that he could arrange the higher payment. The strike was reported in the Western media but not its cause. Anger over the way the relief money was being distributed led to a demonstration on 8 December outside the house of the chief minister of Madhya Pradesh.

The method of determining who was entitled to relief payments has to be read to be believed. One Indian journalist reported:

As for determining who were 'seriously' affected and who 'less seriously', the bureaucrats seemed to have evolved their own strange criteria. This writer met an additional secretary (tribal welfare) at the Hamidia Hospital who was in the process of identifying victims to be given compensation. When asked if he was consulting doctors to enable him to decide the cases, he replied, 'I am looking at their faces and deciding for myself'. Even after having decided in such arbitrary ways who was worthy of getting compensation, the victims were denied their legitimate dues. This writer saw an officer giving compensation (1,000 rupees) due to one of the 'less serious' victims piecemeal – 500 rupees by cheque and 300 rupees by cash; the remaining 200 rupees, he said, could be collected at the victim's convenience, without specifying where it should be collected. (Bhopal – Industrial Genocide?, p. 45)

One other example of corruption, among many, was cited by the filmmaker Ashay Chitre: 'Within the first week, I know of a politician who had managed to get hold of five orphans and registered a society of 65 orphans or something like that, and this was within a week' (APPEN, 1985, p. 35).

Operation Faith

In the aftermath of the killing, the Indian state faced a major practical problem – disposing of the remaining MIC. That this problem was an urgent one was shown when UCIL executives under arrest in the plant refused to leave the plant to be remanded in custody: they claimed the plant was in such a dangerous state that they could not accept responsibility for what would happen if they left it unattended. Similarly Dr S. Varadarajan, of the Council for Scientific and Industrial Research, said that the UCC team from the USA feared the remaining MIC was unstable and argued that it should be processed immediately: 'The chairman of UCC USA sent telex messages asking the government of India to proceed with the processing immediately, as the risk of gas leakage was increasing every day. UCC could not provide specifications for unstable and stable material nor methods for testing' (*Chemical and Engineering News*, 27 May 1985, p. 5). Because of this lack of information, Varadarajan said that he delayed the processing, while making arrangements to increase its safety.

Why did UCC executives show such concern for speedy processing? One reason may be found in the results of analyses of air samples taken on 5–6 December 1984 (that is, three to four days after the killing) by the Pollution Control Board, which found 4.5 ppm cyanide in the air near the MIC storage tank.

This is nearly half the Maximal Allowed Concentration [a 'safe' or 'ceiling' level above which workers may not be exposed to a chemical] and about one-fiftieth of the lethal concentration set for hydrogen cyanide. The cyanide level reduced to half this value fifty metres away from the tank and cyanide was not detected in samples collected from three other localities further away from the factory – the clear inference being that cyanide was still leaking from the MIC storage tank even on the 5th and 6th of December.' (Narayan et al, 1985, p. 10)

Local UCIL management were reported to have brought workers to the plant to process the remaining MIC but were refused entry by the police.

This problem, the *Guardian* (11 December 1984, p. 6) reported, 'is compounded by the antagonism in the city towards the Union Carbide management. Any signs of activity at the plant, which is closed and under tight security, would cause an outcry in Bhopal and the government would be forced to evacuate the entire town – an impossible task.' The state had cause to worry about its loss of social control in Bhopal. Everest (1985, p. 146) reports, 'The day after the killing, several thousand Bhopal residents tried to storm the factory. Plant officials and police guarding the plant, hopelessly outnumbered, only succeeded in turning the crowd away by telling them that another poisonous gas leak was in progress.' On 10 December, local panic was reported at the possibility of the gas being moved. The élite were the first to leave: with reason, they did not trust their own class. One frightened industrialist told the *New York Times*, 'We are getting out today and not coming back until the plant has been cleared of all the gas.'

Operation Faith – the name given by the Indians to the processing of the remaining MIC into the finished pesticide product – fulfilled a social as well as a technical function. It was essential for the Indian state to contain any possibility of disturbance or revolt caused by the killing. The attempted storming of the plant was followed by demonstrations over the distribution of immediate relief. Because of this, some Indian observers argue that the mass exodus from Bhopal during Operation Faith was engineered by a state government anxious to diffuse growing public anger. Thus the people's attempts to organize politically, difficult enough given the state of physical health of many of them, were defused and diffused into an immediate personal concern for safety. This personalized the problem of safety: fight turned to flight and the state demonstrated its ability once again to impose its own life-and-death solutions on its subjects.

While the state wished to avoid a permanent evacuation – such as those following the toxic disasters at Seveso, Love Canal and Times Beach, related to long-term contamination – the short-term evacuation during Operation Faith was acceptable. It delayed the return to 'normality' and the resumption of production that is a major concern of both state and capital in the wake of any disaster, but since Bhopal is mainly a town of the bureaucracy and service industries the impact would be less dramatic. It possessed only two major heavy industries – UCIL itself and the Bharat electrical factory – so production could be temporarily interrupted without major losses.

Given the people's recent experience and their justifiable distrust over the reopening of the factory, some form of evacuation was inevitable. Contradictions in government statements and actions accentuated this. While assuring the public that Operation Faith was 100 percent safe, it closed schools and provided extra transportation out of the city. It also set up special camps in that part of the city farthest from the factory for those who were unable to leave the city. In a typical move, chief minister Arjun Singh told the public that he would be in the plant during the processing to show how safe the process was and to show his faith in Indian science and technology. Not many citizens were impressed with this: one observed to the *New York Times* (14 December 1984, p. A10) that the chief minister 'has a car, a helicopter, buses, trains in which to get away – we have nothing'. Singh did not limit himself to material means of assistance only:

he called on the people of Bhopal to pray for him and for the success of Operation Faith.

Although the mass media characterized the local people's reaction to Operation Faith as one of 'panic', it is important to point out that there was a sound scientific basis to their fears. One of the characteristics of MIC is that it sensitizes the exposed person, with the result that further exposure, even 'mild', can result in allergic reactions which, in some cases, may be lethal. Furthermore, despite the Indian state's propaganda that it would be in control of Operation Faith, it was obvious that Union Carbide management and staff were not only running the factory during Operation Faith but were also calling the shots as to how the remaining MIC was to be disposed of.

The choice of technical means involved was dictated by UCC's economic interest.

Operation Faith may be seen as a victory for the balance-sheet view. UCC was pushing hard for the remaining MIC to be processed into finished pesticide form and had immediately recommended its units throughout the world to use up remaining MIC stocks before governments moved in to close the plants. One possibility was that the MIC would be returned to the parent company for neutralization. UCC very much wished to avoid this; as it already had one consignment of MIC on the way back to the USA, it didn't want any more. Quite a large proportion of the charge[2] UCC took in its fourth-quarter results against Bhopal related to transportation and neutralization of MIC, along with the consequent loss of finished product. Cost–benefit analysis ruled in Operation Faith. It would have been possible, for example, to neutralize the remaining MIC under a nitrogen blanket using caustic soda, but this would have been all cost and no benefit to UCC.[3]

Operation Faith was unquestionably a combined production of the state and capital. Many observers have condemned it as a further example of the Indian state's kotowing to UCC's priorities and said it exemplified the technological poverty of the Indian state. This view needs to be qualifed by part of a report of a detailed study by the Union Ministry of Chemicals and Fertilizers in February 1985; the report said UCC refused to provide information to the Indian state after the Central Bureau of Investigation seized the Bhopal plant, including information on how to dispose of the remaining MIC (*Economic and Political Weekly*, 16 February 1985, p. 254). Thus UCC may have forced the Indian government into accepting UCC's control of the decision on how to dispose of the remaining MIC. This required participation by UCC experts from the USA, who had previously been refused entry to the plant for fear they might destroy evidence. The CSIR's Dr S. Varadarajan, the chief Indian government scientist in charge of Operation Faith, told the Indian Science Congress in January 1985 that his team had found no one at Bhopal who had any idea of the chemistry of MIC. Engineers at the plant went by the operating manual only and did not understand its design.

Nevertheless, the Indian state put the best possible face upon Operation Faith. It was described as '100 percent' safe and was 'supervised' by fifty Indian scientists. With the double-speak normal for such a situation, the Indian government described the pesticide production process as a 'neutralization' process. By the time Dr Varadarajan finally finished his report on Bhopal in December 1985, the part played by UCC staff had been forgotten: Varadarajan's report omits any mention of UCC staff involvement, giving the impression that Operation Faith was totally conceived and controlled by Indian government scientists and technologists.[4] At the time, however, Varadarajan told the *New York Times* (21 December 1984, p. A9), 'All the work has been done by Union Carbide people under the supervision of our scientists . . . No one else is in charge. Nobody can operate the factory except the factory manager and his staff. The scientists were monitoring and supervising the process.'

The amount of trust that should be placed in these UCC experts was shown when it was reported that the amount of MIC processed during Operation Faith was 22 tonnes, nearly 50 percent more than the UCC experts had estimated remained in the plant. 'So much for scientific expert caution and scientific exactitude', commented the local environmental group, Eklavya. Some of this extra MIC can be accounted for by the 1 tonne of MIC transferred to another tank in October 1984. Plant officials told Indian government officials that they had forgotten about it.

Operation Faith was successful for both UCC and the Indian state. As well as disposing of the remaining MIC, it helped the state re-establish social control, broke up communities and scattered people from Bhopal throughout Madhya Pradesh State. Some of them returned to their own villages and did not return to Bhopal until much later, if they returned at all.

An interesting parallel can be drawn here with the crisis management of Three Mile Island. By avoiding the spectacular threat of a meltdown at the Harrisburg nuclear plant, the nuclear industry could claim the Three Mile Island incident demonstrated the 'safety' of nuclear power. The industry was able to claim a victory over a spectacular possibility, rather than admit the defeat involved in the actual release of radiation by the accident. Similarly, Operation Faith served to hide the threat that the people of Bhopal faced from the slow death that was attacking some of those exposed; it focused attention on the spectacular possibility of another leak. Thus the fear of further acute exposure to MIC was raised, while the chronic effects of the original exposure were minimized and denied.

After Operation Faith

After Operation Faith the Indian government proclaimed that the crisis was over. Following the safe disposal of the remaining MIC as a catharsis, everything should now return to normal. Thus when Dr Varadarajan was asked on 20 December 1984 what other hazardous chemicals were present in the Bhopal plant, he mentioned chlorine and alpha naphthol, ignoring the presence of other toxic chemicals such as carbon tetrachloride, chloroform, monomethylamine and phosgene. In keeping with the Indian government's refusal to provide information, he refused to give details on the quantity of toxic chemicals still stored at the plant. Similarly, the state gave no information on what would be done with the Bhopal plant. The only certainty was that UCIL would no longer be allowed to operate it. Continued unexplained activity at the plant led people to fear another leak, not without cause.

On 22 March 1985 the government began to move chemicals such as chlorine and monomethylamine out of the factory. Again, secrecy was paramount. The *Hindustan Times* (23 March 1985) reported: 'The police and the security guards of the factory nearly pounced on some newsmen yesterday as they were trying to take pictures of the tankers positioned in the factory and as the tankers moved out of the plant. The police objected to the media men watching the mysterious shifting of these materials even from outside the boundary wall of the factory.' During these transfer operations, chlorine leaked on 28 March and three workers were treated for exposure. On 1 April 1985, chlorosulphuric acid fumes leaked at the plant. Hundreds of slum-dwellers fled the area. On 2 April, Motilal Vora, the new chief minister of Madhya Pradesh State, told the state assembly, 'There is no toxic chemical left in the plant and whatever other chemical is being stocked is being removed from the plant' (*Guardian*, 3 April 1985, p. 9). Later the same day, however, a UCIL vice-president confirmed chlorine had leaked but said, 'It was a minor matter. No one was affected and no one was hospitalized' (*Financial Times*, 3 April 1985, p. 4).

In the period after Operation Faith, the Indian state believed a quick deal with UCC was on the cards. The state, of course, wished for a return to normality, business as usual, and also wished to minimize the expense involved in post-disaster work. The prospect of an imminent deal removed the need to discover the causes of the killing; therefore the government crawled at a very slow pace in its public and private examination of the leak's causes.

Business India (25 February 1985) reported that a growing suspicion existed that the results of the official investigation would never be made public. The criminal investigation by the Central Bureau of Investigation, which was originally supposed to be completed within a month, was reported to have barely begun. The Madhya Pradesh Commission of Inquiry, which was supposed to have reported by 15 March 1985, did not hold its first meeting until the end of March. With the Central Bureau of Investigation put on the back-burner and the Commission of Inquiry obstructed by the local state, happily enough the details of the Indian politicians' complicity with UCC would be covered up. In April 1985 the Central Bureau of Investigation report was expected to be completed by the end of May (*Sunday*, 7–13 April 1985). It has yet to appear.[5] In keeping with its cover-up activities, the Indian government proposed destroying the remaining MIC from the tank which had leaked, thus making its analysis and testing impossible. This move was stopped only after a High Court action forced preservation of 15 kilograms for examination (Everest, p. 150). This policy also entailed a major medical cover-up (see Chapter 3).

The government paid little attention to the need for relief and rehabilitation, as it expected the quick arrival of US dollars. Similarly, the prospect of a negotiated settlement removed the need for studies of the gas's effects and the number of the population affected, which would be essential for a court-room battle with UCC. The situation was complicated by the continued existence of a pro-UCC power bloc inside the Indian state scientific/medical bureaucracies. UCC's continuing influence was demonstrated when local UCIL management arranged for the Central Bureau of Investigation to detain and question for two days a production crew that had filmed the Bhopal plant for the BBC television programme *World in Action* (*Listener*, 6 June 1985, p. 7).[6] The government's interest in ending the crisis coincided with UCC's desire to deny that the killing had long-term effects. Faced with this coalition of interests between local state and international capital, various struggles arose around the right to know, medical treatment, the future of the Bhopal plant and the issues of relief and rehabilitation.

The first struggle took place over state provision of free rations. Before the elections took place, the local government made free rations available to the majority of the population. Though only some 250,000 people were affected by the gas, some 700,000 fresh ration cards were issued by Arjun Singh's state government. This was transparently a political ploy to gain support for the forthcoming elections. In January, following the elections, the government halted the free distribution of rations, giving as its reason the need to investigate who was entitled to get relief. The various voluntary organizations, which had cast the Bhopal killing in the mould of a hurricane or some other 'natural disaster' that required only interim relief, also pulled out of Bhopal.

The people protested at the cutting off of relief. On 1 January 1985 the Nagrik Rahat aur Punarwas Committee mobilized a *chakkajam* (Stop the Wheels) protest by organizing people to squat on Bhopal's main thoroughfares. On 3 January, some 10,000 people took part in a protest march. The same day the Zahreeli Gas Kand Sangharsh Morcha (Poisonous Gas Disaster Struggle Front) organized a 'Dhikkar Diwas' (Day of Condemnation) and a *dharna* (relay fast) in front of the residence of Arjun Singh, chief minister of Madhya Pradesh state. This developed into a ten-day *dharna*, and finished with a 'Rail Roko' (Stop the Train Wheels) protest.

The state didn't take this lying down: ten people were hospitalized on the second day of the *dharna*, having been beaten by the police when the protesters had threatened to storm the chief minister's residence. Dr Sadgopal and seven other Morcha activists were arrested at the chief minister's house. The police rounded up other Morcha activists before they reached the railway station and held them. At the railway tracks some 300 people were arrested, at least half of them women, held overnight and released without being charged the next day. Dr Sadgopal and the seven other fasters from the Morcha were told they would be released if they signed an undertaking to stop organizing in the slums. They refused, were held for eight days and were released without being charged. Despite this repression, the rations were restored, though the state constantly threatened to cut them off again.

The composition of the rations also became an issue, not without cause. Praful Bidwai, reporting for the *Times of India*, noted that

a quarter of the free rations recorded as being doled out was not reaching the people, while the quality of the rations was very bad: in some cases as much as a third or four-fifths of the wheat is stones, pebbles and other muck, and the milk is liberally adulterated with water and finally, several thousand poor families who desperately need support do not receive any free rations because they have not been issued ration cards (as they live on unauthorized plots of land) or have been tricked by local ration-shop owners and petty officials into surrendering them. (27 March 1985)

As well as responding to protests with repression, the state sought to divide the affected community to prevent it from uniting in response to the killing. The distribution of free rations contributed to the increasing tensions between the various areas of Bhopal. A doctor from Bhopal told *Comhlamh News*:

A few days after the disaster the streets of the slums nearest to the Union Carbide factory were thronged with various aid agencies distributing food, blankets, clothing, medicine etc. to people who lived in the area. 'It was like a race to see who was most generous – who could give most things away. Families who had not suffered received the same as families who had.' In response to this many of the men stopped working, got a certificate from a

One of the many demonstrations organized by the Morcha. This one, in January 1985, demanded 'No reopening of the plant', in response to frequent rumours that it would resume operations.

local doctor to state that they were affected by the gas and joined the handout queues. Meanwhile inhabitants of the other areas of town, unaffected by the leakage from the plant, began to resent the growing level of prosperity enjoyed by those who lived in the disaster area. This helped to cause antagonism and tension and brought about a split in the community. On several occasions lorries carrying supplies to the affected areas were stoned by other residents and in some instances the contents were seized. (Autumn 1985, p. 28)

This doctor's evidence has to be treated with some suspicion, given his claim about the easy availability of certificates stating persons were affected by the gas; all the other coverage implied it was extremely difficult to obtain such certificates. Nevertheless, the description of splits in the community rings true.

The government also went out of its way to exploit the division between the Bhopal plant workers and the slum dwellers. It attempted to frighten the workers away from organizing with the gas victims by saying the victims held the workers responsible for the killing. 'When workers raise demands, they are reminded of the worse fate of the gas victims. Thus the government uses the gas victims to put down workers' demands' (*Economic and Political Weekly*, 14 December 1985, p. 2200). The government also wished to disperse the plant workers. It set up an employment exchange specifically for UCIL workers, following its announcement that the factory at Bhopal would not be reopened. The government undertook to provide alternative employment for the workers, who had thereby lost their jobs. This promise was as shabbily carried out as all the other government promises: workers were offered jobs far from Bhopal, generally at lesser rates of pay which did not take account of the workers' skills and experience.

Organized Opposition

The response from organized opposition forces in India varied widely. The following account is necessarily sketchy. More detailed analysis of opposition forces around Bhopal will, one hopes, be provided by the people involved, as lessons from Bhopal would help people opposing toxic industry elsewhere.

The response by the capitalist opposition parties was opportunistic and sporadic. Attempts by these parties to make political capital out of the Bhopal killing were not highly successful, given the fact that they shared the same attitude and some of the same relationships with toxic capital as did the party in power, Rajiv Gandhi's Congress (I) Party. The response by the mainstream left was dismal. Praful Bidwai, assistant editor of the *Times of India*, has pointed out that one reason for the inadequate relief and rehabilitation measures provided by the Indian state was the weakness of the pressure exerted by opposition forces. Just as the Indian state treated the killing not as a national but as a local disaster, so also the opposition forces within India did not organize adequately on a national basis on the issue of Bhopal. 'The primary cause of this weakness is the lack of response to this issue by the trade unions and the communist parties' (*Inside Asia*, November 1985, p. 42).

This lack of response may be illustrated anecdotally. In September 1985, Suraj Patil, a minister of state, said in the Lok Sabha (the national parliament) that UCC's Bhopal research and development centre's operational licence had not been renewed. However, no licence is needed for R&D activities. A Morcha activist showed a West Bengal leftist MP documentary proof and asked him to move a motion accusing the minister of misleading parliament. The MP defended Union Carbide, said parliament should not be bothered with such trivial matters and told the Morcha not to overestimate the importance of Bhopal. This is just one example of the many complaints by activists that opposition political parties and mass fronts, particularly the leftists, were not concerned with Bhopal and the issues it raises.

This failure to organize on the issues Bhopal raised is to be expected from the communist parties and trade unions of India. Both of the Indian communist parties view their natural base as among the industrial working class and their task to be the organization of this class. Given that the continued growth of both communist parties is based on the continued growth of industrialization in India, they are unlikely to either welcome or articulate criticism of this industrialization. Indeed, the party in power in West Bengal, the Communist Party (Marxist), is involved in attracting multinational capital and increasing state control of the working class: 'Seeking to assert itself as a better manager of regional interests than the Congress, [the Communist Party (Marxist)] has not surprisingly betrayed its election pledges by tightening controls on the West Bengal labour force, as part of a drive to attract public and private (including multinational) investment to the region' (Vanaik, 1986, p. 60). This uncritical attitude to the forces of production is a characteristic of many socialist and communist parties.

Opposition groups that have organized in and on Bhopal mainly come from the autonomous sector, which has grown recently in India. Generally these groups have consisted of independent non-party leftists and radicals, with some input by Marxist-Leninist groups. Activities have divided into two areas, research and organization, though opting for one area has not excluded participation or support for the other.

In the former, work has been done by environmental groups such as Eklavya, radical science groups such as the Delhi Science Forum and radical health groups such as the Medico-Friend Circle. Both the Bhopal-based Eklavya and the Delhi Science Forum produced excellent English-language reports and circulated them within a short time after the killing. The first version of the Delhi Science Forum report, for instance, was presented to the press on 18 December 1984. While one might take issue with various aspects of the Delhi Science Forum report in particular, the production of these reports was a commendable example of a speedy response by a radical science group to a major popular issue. The quick production of these reports also influenced the media, given that no such accounts and investigations were made available either by UCC or the Indian state. However, it must be noted that most of these reports and analyses were produced for the intellectual élite: there appears to have been little attempt to communicate their findings to the gas-affected people.

Radical health groups such as the Medico-Friend Circle occupy an intermediate position between the poles of organization and research. Thus, as well as doing excellent medical epidemiological research, the Medico-Friend Circle set up health projects. Along with other radical health groups such as Drug Action Forum, it provided the medical staff for the People's Health Clinic, set up by the combined opposition forces in June 1985. A similar position, half-way between research and action, was occupied by the Bombay-based trade-union research organization, the Union Research Group. Finally, action in its own professional area, as well as research, was undertaken by the Bombay-based Lawyers' Collective, a leftist professional group which made their professional skills available.

Two major groups concentrated on organizing opposition on the ground in Bhopal. These have already been mentioned: the Nagrik Rahat aur Punarwas Committee, associated with the film-maker Tapan Bose, and the Zahreeli Gas Kand Sangharsh Morcha, associated with Dr Anil Sadgopal. Women's groups also took part in both organization and research, in particular the Delhi-based women's group, Saheli. The importance of women in the struggles after the killing cannot be overestimated: women activists were also prominent in many of the mixed groups, and the women of Bhopal themselves were extremely prominent in the protest activities.

The leftist opposition in Bhopal was plagued by the normal problems familiar to anyone anywhere with experience of the radical left. These included jockeying for power inside popular organizations and fronts, parachuting ready-made programmes on popular issues and movements, autocratic organizations and hierarchical structures, concentration on ideological differences rather than practical co-operation, dogmatism, the desire to be seen as *the* group representing the victims, and personality clashes. While organization and protest by leftist-led groups attracted a large number of gas victims initially, one Indian comrade reports this 'gradually tapered down to a dismal state due to a combination of lack of effort towards politicization of cadres/masses, autocracy within the organization, hopelessness among the people setting in fast due to lack of any history of organized struggle and of course the state's repression'.[7]

Attempts were also made to organize on a national basis. On 17 and 18 February 1985, a national convention in Bhopal was attended by 150 delegates from about sixty-five organizations in thirteen different Indian states. This convention saw the formation of a Rashtriya Abhiyan Samiti (National Campaign Committee), whose aims were to campaign on a subcontinental scale on the issues raised by Bhopal and to support local organization and relief efforts in Bhopal. The Committee organized national protests and supported groups in Bhopal both politically and financially. Disagreements and disputes also arose between some groups in the National Committee: in one case, for example, the Delhi Committee on the Bhopal Gas Tragedy was asked to withdraw its pamphlet, *Repression and Apathy in Bhopal*. Similar support for the workers' struggle and for general rehabilitation efforts came from the formation of the Trade Union Relief Fund.

International support and solidarity was also organized. Speakers from India toured communities in the USA where UCC plants are located. Other activists visited Britain, Ireland and other European countries. In Britain a conference on the issues raised by Bhopal, which was attended by Indian activists, was organized in November 1985. Protests at various UCC offices also took place. In the USA a conference was organized for two days in March 1985 at Newark, New Jersey, on the theme 'After Bhopal: Implications for Developed and Developing Nations'. There were also expressions of solidarity in Asia, particularly through the Asian Pacific Peoples' Environment Network (APPEN), which also produced an excellent publication making available basic documents relating to the killing and the opposition. Hong Kong's Arena Press also quickly produced an excellent selection of articles on the killing from Indian publications.

One excellent trend following Bhopal was a growth in co-operative research between groups in peripheral and metropolitan countries on Bhopal and its implications. The Highlander Research Center, in New Market, Tennessee, USA, co-operated with the Society for Participatory Research in Asia in producing a document dealing with Bhopal and its implications. Ward Morehouse of the US Citizens' Commission on Bhopal co-operated with M. Arun Subramaniam of the journal *Business India* to produce a report. Environmental researcher Barry Castleman co-operated with Prabir Purkayastha of the Delhi Science Forum to produce an excellent case study of double standards between UCC's plants at Bhopal and Institute, West Virginia. These efforts were useful in helping spread information on what happened at Bhopal and its implications.

Practical support was more difficult to organize. The struggles in Bhopal after the disaster developed on two fronts. The first related to relief and rehabilitation of the gas victims and concentrated on the medical treatment of the victims. The second front focused on a possible alliance between the

plant workers and the gas victims, which grew out of the workers' demands for alternative employment and eventually developed into an unsuccessful demand for alternative production.

Workers' Struggle

The struggle of UCIL workers in Bhopal is the one that has received most support from the Western left, as well as most coverage in its press. This struggle has been presented as a unified one shared by both workers and gas victims, as an example of a struggle for conversion of toxic industry and even as the beginning of a case for workers' control of industry. The workers' struggle arose from the Madhya Pradesh state government's avowed aim of ensuring the Bhopal plant would never reopen under UCIL. In the aftermath of the killing, UCC has offered to open a battery plant in Bhopal but this was rejected out of hand by the local state. Interestingly, however, the R&D centre was allowed to continue operating, though its tax-free status was suspended. In April 1985 UCIL announced the factory would be closed in the beginning of July.

One of the workers' trade unions began a campaign calling for the preservation of the workers' jobs or the provision of alternative employment. This campaign received support from the Bombay trade unions and from the Bombay-based Union Research Group. The basic campaign for employment had another demand added to it: that alternative production should take place in the UCIL plant in Bhopal, which would provide jobs not only for the now unemployed chemical plant workers but also for a proportion of the gas victims. Furthermore, the products to be made by this alternative production should be socially useful. Capping all this was the presumption that this alternative production would be managed by the workers. On 14 June 1985, 400 people stormed into the plant to begin a sit-in protest over job losses. Workers, with the co-operation of the Union Research Group and some local academics, proposed that the UCIL plant be converted to produce soya-based food products and agricultural machinery. What had started as a traditional struggle by the representatives of variable capital (labour power) for jobs and compensation was thus given a new twist.

This campaign was criticized on various grounds (Jackson *et al.*, 1985, pp. 11–12). The main idea of alternative production was imported from the Bombay trade unions. The assumption of the Bombay activists that the plant would operate under workers' control overestimated the sophistication of the Bhopal workers. Bhopal lacked a tradition of trade-union struggle, which might possibly have laid the foundation for workers' control; Bhopal is basically unindustrialized, like the state of Madhya Pradesh in general. The technical practicality of alternative production was also questioned by trade unionist Jacky Vidal from the UCC plant at Béziers in France, who had been a member of the international trade-union investigatory mission to Bhopal. Finally, of course, there were grave doubts as to the suitability of producing food products in a plant which had previously been used to process highly toxic chemicals. One of the doctors arrested in the government repression in Bhopal in June 1985, Nishit Vora, commented: 'So this whole idea of alternate production, as far as I'm personally concerned and to many others who are basically more active with the gas victims, is not feasible either socially or technically' (Jackson et al., 1985, p. 12).

Despite credit due to the opposition in Bhopal, these criticisms seem warranted. It also appears that the much-heralded alliance between gas workers and gas victims was more illusory than real. The state repression at the end of June 1985 which resulted in the arrest of many activists and the closure of the People's Health Clinic passed the trade unionists by: their leaders were not arrested in the midnight swoop. This exemption was reportedly due to a deal between the police and the trade unions in which, in return for immunity from arrest, the unions agreed not to take part in the forthcoming demonstration. Regardless of whether or not such reports were true, the tenuous 'solidarity' between workers and gas victims was broken.

On another level, it was unrealistic to expect the trade unions to be concerned about anything other than the sectional interests of their members. The purely sectional character of the trade-union demands has been condemned by Indian activists: 'These unions, in common with most of the unions in other Carbide plants, have conspicuously failed to launch any agitation or industrial action to support the claims of the gas victims' (*Economic and Political Weekly*, 14 December 1985, p. 2198). The trade unions were also quick to abandon the alternative production plans.

Originally the Union Carbide Karamchari Sangh, the trade union not affiliated to Rajiv Gandhi's Congress (I) Party, had demanded alternative production at the Bhopal plant. When the plant was closed, this demand was dropped in favour of alternative employment and higher redundancy payments. These demands were identical to those of the union affiliated to the Congress (I) Party, the Union Carbide Workers' Union.

The workers' occupation of the plant created major problems for UCIL's plans to sell off the plant. Yet for public-relations reasons it was not possible for UCIL to ask the state to remove the occupying workers. UCIL tolerated the problem, as it hoped resolution of the compensation issue regarding jobs and gratuities would end the occupation. This eventually happened in December 1985, when UCIL made a large cash settlement with the workers in return for an end to the factory occupation and an end to any further demonstrations against the company by the workers.

Repression and Inadequate Rehabilitation

In accounts of repression, most attention has been paid to the closing down by the police of the People's Health Clinic in June 1985. That same month the police also closed down a women's centre which had been set up in Bhopal with food, library and other support facilities. The repression in June was an attack not only on the health activities of the activists but also on the accompanying agitation. Activist groups in Bhopal had realized from the beginning that the problems involved in Bhopal were too great for non-governmental organizations to cope with. Thus rehabilitation activities were constantly coupled with attempts to organize the gas victims to demand their basic rights to adequate relief and rehabilitation from the government. The repression in June 1985 was carried out both by the police and by local political activists and thugs associated with Rajiv Gandhi's Congress (I) Party.

The mass arrest on 24 June, with the closure of the People's Health Clinic, was intended to prevent the protest rally due to take place the following day. Up to forty people were arrested by the police, of whom only six were doctors; the rest were activists of the Morcha and the Nagrik Rahat aur Punarwas Committee, as well as activist slum dwellers. While the police claimed that activists were arrested to prevent trouble at the following day's rally, which had been organized by the Morcha, the real intent was to try to prevent the rally from taking place and thus impede the developing struggle of the gas victims.

This policy of arresting presumed leaders promoted the state's claim that outside agitators alone were responsible for the protests in Bhopal. On the night of 24 June, Congress (I) Party goons put posters on the walls of the UCIL factory alleging that the Morcha had CIA links. These posters were also distributed among the slums. Morcha activists claim that Sudip Bannerjee, former director of the Madhya Pradesh state information and publicity department, told press reporters that the Morcha had links with UCC. The Madhya Pradesh governor called Morcha activists 'professional agitators and vested interests', while the chief minister branded them as outside elements. Accusations on posters have claimed Morcha activists were CIA agents or UCC supporters. Consistent with this attempted criminalization of dissent, the Morcha offices were placed under surveillance. The Local Intelligence Bureau asked the shop where the Morcha did its photocopying to make a copy for them of all Morcha material.

The arrest of these 'outside elements' did not prevent the rally, which was intended to take place outside the Madhya Pradesh government secretariat: some 3,000 to 4,000 people attended it. A police cordon had been thrown around the secretariat building. Even the protest was manipulated. When the demonstrators tried peacefully to pass through the police cordon and *gherao* (obstruct and occupy) the secretariat, the police charged the demonstrators on the pretext that stones were thrown at them by the crowd. The stone-throwing was done not by the demonstrators but by Congress (I) Party goons who had infiltrated the crowd. Some thirty-one activists were arrested after the police charge. The state's attempt to criminalize dissent and protest required that demonstrators be smeared with a violent tag. As well as the organized stone-throwing, two activists – arrested on 24 June and illegally confined by the police until the 27th – were charged with approaching the UCIL factory on the 25th at 10.30 a.m. with cans of kerosene. Many of the charges would not succeed of course, yet they succeeded in impeding agitation, both by diverting activists to defending themselves in the legal process and by smearing the protesters with the violence planned by the state's allies.[8]

The state continued its policy of inaction on relief and rehabilitation combined with repression of dissent and protest. In July 1985 the London *Times* reported (11 July 1985, p. 1) that the state government had not carried out a survey of those incapacitated and requiring sickness benefit, while only 906 families of those who had died had received the £605 (10,000 rupees) state grant. On the medical side, according to one doctor, the pace of detoxification injections for the 110,000 people affected was going so slowly that it would take seven years to complete. In May 1985 a large number of fish died overnight in Bhopal. Water was sent for testing but no results were made public. This inactivity was normal. The previous March, Professor S. K. Roy and and D. S. Tripathi of Banaras Hindu University had said tests on crops 'showed that MIC acted as a mutagen and changed the morphology and breeding behaviour of the plants . . . They advocated destruction of all standing crops and keeping the land fallow till after the monsoons' (*Madhya Pradesh Chronicle*, 16 March 1985). Needless to say, this was not done. Similarly, the Nagrik Rahat aur Punarwas Committee reported that the water in Bhopal was polluted, but no action followed.

Investigations into the killing ground to a halt. The Central Bureau of Invesitgation report did not appear. In September 1985 the Madhya Pradesh government indicated that the state Commission of Inquiry might be suspended. The offical reason for this was that it did not want to prejudice the claims against UCC in the USA. The real reason was that the Commission might further expose government complicity in the events leading up to the killing: the Commission had already brought to light material that was embarrassing not only for UCC but also for the Indian state. The Commissioner had also publicly complained of utter non-co-operation on the part of the government.

In December the report to the government by Dr Varadarajan was finally made public. What was most interesting about the Varadarajan report was how it mirrored in many respects the UCC report of the previous March. As *Chemical and Engineering News* noted (6 January 1986), 'The basic chemistry in the Indian report largely agrees with Carbide's findings last March.'

There were two areas of disagreement in the chemical analysis of the residue: the first related to the percentage of metal contaminants in the residue; for the second, the Varadarajan report (1985, p. 31) notes, 'the residue samples contain about 3–6 percent of unidentified tarry materials'.

Where the two reports differ is in regard to failures in design and operation of the plant. Varadarajan (p. 81) concludes:

The needless storage of large quantities of the material in very large size containers for inordinately long periods as well as insufficient caution in design, in choice of materials of construction and in provision of measuring and alarm instruments, together with the inadequate controls on systems of storage and on quality of stored materials as well as lack of necessary facilities for quick, effective disposal of material exhibiting instability, led to the accident. These factors contributed to guidelines and practices in operations and maintenance. Thus the combination of conditions for the accident were inherent and extant.

The Varadarajan report mirrors UCC's in another important way. It also refused to provide information on the composition of the gas cloud.

Since simulation of the exact conditions of the total event that occurred in tank 610 was not possible, experiments were carried out by taking small quantities of MIC and subjecting them to different reaction conditions . . . Gaseous products have not been examined in these experiments. Further work will be done to identify and estimate the gaseous products as well. (Varadarajan, p. 55)

None of the results of this further work had appeared a year after the report's publication. Nevertheless the report estimates that the escaping material included 80 kilograms of ammonia and 1.25 tonnes of carbon dioxide. Varadarajan also deals with the cyanide breakdown possibility by suggesting (p. 60), 'It appears chloroform inhibits the breakdown of MIC to hydrogen cyanide.'

The Varadarajan report continues the chorus of reassurance regarding pollution. It states (p. 14) that Indian government scientific 'teams made an examination of the environment immediately to ascertain if the presence of any isocyanate or other related materials could be detected in air, water or surfaces'. The results were reassuring: 'The tests carried out did not show any presence of MIC or related toxic materials in the environment' (p. 14). This handily ignores the report of the Central Water and Air Pollution Control Board, which found cyanide at a level of 4,533 micrograms per cubic metre inside the UCIL plant near the MIC storage tank and at a level of 2,533 micrograms per cubic metre fifty metres from the MIC storage tank on 5–6 December 1984 (APPEN, 1985, p. 118). It also ignores the report by the Nagrik Rahat aur Punarwas Committee, released on 2 May 1985, which found a high level of thiocyanate in the subsoil lakes and filtered water of Bhopal, even 100 days after the killing.

Nor was the state any more forthcoming with information. Journalist Sujit K. Das complained: 'Though socio-economic and health surveys of about 80,000 people and 25,000 families have been conducted by the Social Welfare Department and the Tata Institute of Social Sciences respectively, the reports have been held back' (*Economic and Political Weekly*, 14 December 1985, p. 2192).

Meanwhile, actual relief and rehabilitation schemes staggered on. Little attempt was made to provide alternative training and employment for those unable to return to their former work due to the gas leak's effects. The thought and care devoted to such alternative work schemes as were set up can be seen in the fact that the major work made available by the state to the women gas victims was sewing – making victims use their eyes, which had already been badly affected by the gas leak. In August 1985, the Madhya Pradesh government proposed spending 3,320 million rupees for the rehabilitation of the gas victims and development of the gas-affected areas. Chief minister Motilal Vora hailed the proposal as a significant rehabilitation measure. It was in fact an example of pure opportunism on the part of the state. Of the 3,320 million rupees, only 380 million rupees were intended for schemes related to the gas victims. The rest was to be spent on construction of an airport, modernization of urban transport and the railway station in Bhopal and beautification of the city.

On 4 November 1985, an independent committee was appointed by the Indian Supreme Court to oversee quick medical relief and compensation. This committee was composed of two representatives of the Indian Council on Medical Research and the Madhya Pradesh government, Dr Heeresh Chandra, Anil Sadgopal from the Morcha and a member of another voluntary organization. Nothing further has been heard of this committee. Given the Madhya Pradesh government's response to the Indian Supreme Court's demand in July 1985 for the production of a time-bound detoxification scheme, this is no surprise: by November 1985, the state government had still not submitted this scheme.

Despite grandiose plans and schemes, the position of the gas victims was as bad as ever. The *American Lawyer* reported:

> *Moreover, people who have been to Bhopal recently say they see few signs that people are getting as much assistance as the government and private relief agencies claim to be providing. For example, last Christmas Eve a group in California led by Dr Vin Sawhney, a gastroenterologist, contributed $25,000 to the Indian Red Cross to set up mobile medical units in Bhopal. The units were needed, Dr Sawhney believed, because most of the victims lived far from the centre of Bhopal in squalid shanty towns. In March a member of the group, Vivek Pinto, visited Bhopal to see how the money was being used. 'Nowhere did I see the two mobile medical units for which a sum of $25,000 was donated', he reported in a letter to Dr Sawhney. 'What disturbed me much more was the utter placidity of the [Red Cross] office and its personnel. It was as if nothing had ever happened in Bhopal.'*
>
> *In July Dr Sawhney visited Bhopal himself. 'At that time they still hadn't had [the mobile units],' he says. 'They said they would be functioning in two or three weeks. I asked them to write to me and they haven't yet.' He adds, 'We were planning to send more money, but after that we decided not to.' (November 1985, p. 58)*

If relief and rehabilitation were lacking, repression was not. In September 1986 two members of the Bhopal Group for Information and Action and one member of Suraksha were charged with violating the Official Secrets Act. The British member of Suraksha was also denied a visa extension. Suraksha was a relief project working mainly with the children of the shanty towns. With the assistance of local artists, it was attempting through painting, music, story-telling and drama to help the children deal with the psychological trauma the leak produced. The Bhopal Group for Information and Action had set up a documentation centre and was researching and publicizing various aspects of the killing. It tried to keep the issue of Bhopal alive through the publication of a monthly newsletter.

When the repression began, the Group was researching UCC's attempts in Bhopal to substantiate the firm's sabotage theories; then the police seized most of the Group's files.

The official secrets charges were for tape recording a public meeting on the medical condition of the gas victims, at which differences between

WE WILL NEVER

On the SECOND ANNIVERSARY of the world's worst industrial disaster, let us pledge to fight multinational interests and governmental apathy to ensure that there shall be NO MORE BHOPALS

NOV/DEC 1986   ISSUE 6 & 7   Contributory Price Rs 5/-

>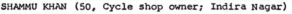
> SHAMMU KHAN (50, Cycle shop owner; Indira Nagar)
>
> People are still going around in circles for their 1500 rupees relief money. The assets of Carbide are still intact. Neither is the government taking it over, nor is it using Carbide's assets to help the poor victims. The people are not quiet, it's just that they are being lulled. Like when a child cries, one soothes it by diverting its attention saying a tiger is coming or a goat is coming. Neither does the tiger come nor does the goat. And the child eventually sleeps. The government is working in a similar fashion. We will have to cry out all over again.
>
>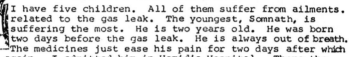
> KANTI BAI (30, Housework; Vijay Nagar)
>
> I have five children. All of them suffer from ailments related to the gas leak. The youngest, Somnath, is suffering the most. He is two years old. He was born two days before the gas leak. He is always out of breath. The medicines just ease his pain for two days after which it comes back again. I admitted him in Hamidia Hospital. There the doctors said he had TB. He was given three injections a day for eight days and then he was discharged. After some time he became just as sick again. Before the gas no one in my family had suffered from TB.

doctors over the treatment of the gas victims were discussed. This was, of course, only a convenient excuse for the repression. David Bergman, the English member of Suraksha, explained the reasons behind his arrest to the *Guardian* (31 October 1986, p. 8): 'In a situation where voluntary organizations are the only people attempting serious rehabilitation projects, people who would like to be involved are too scared. Any relief effort which the government cannot control is seen as a threat. It highlights their inadequacy in failing to solve the physical and mental health problems of the gas victims.'

The local Bhopal paper ran smear stories accusing the arrested activists of being agents for UCC. Activists responded to the arrests by a series of press conferences exposing the true situation in Bhopal. They staged a sit-in to protest at the arrests outside the Madhya Pradesh government offices in New Delhi, and they sent the Indian Supreme Court a petition calling for freedom from harassment and for the right to information. In an apparently unrelated protest, some 2,000 people, including 500 women, were arrested on 29 September 1986 in Bhopal during a protest to demand more aid for the gas victims.[9]

By its response to Bhopal, then, the Indian state starkly revealed itself as more concerned with protecting its image and its economic policies than with protecting its citizens. It repressed citizens' initiatives, monopolized and censored information and overall was more interested in managing the crisis than in dealing adequately with the health effects through relief and rehabilitation measures.

3 THE MEDICAL COVER-UP

In the immediate aftermath of Operation Faith, the Indian government maintained its line that MIC has no long-term effects. Tapan Bose, of the Nagrik Rahat aur Punarwas Committee, wrote to the *Economic and Political Weekly* (12 January 1985, p. 46) 'The Ishwar Dass Committee [chaired by Dr Dass, commissioner in charge of relief of gas victims] has declared that these patients who are still reporting for treatment are not the victims of MIC exposure but patients suffering from chronic disorders.' Journalist Ivan Fera reported that when people returned to the hospitals with new symptoms, the hospitals no longer recorded on their prescription forms that they were gas victims.

One major practical reason the state had for denying the extent of the killing was the cost, likely to be immense, of medical assistance to the exposed population. This cost, which was high enough already, would be enormous if the government were forced to provide lifelong health services to the gas victims.

In a situation where the state already has immense difficulties providing basic medical and health facilities for its population, the health services in Bhopal were grossly overstretched, having already been overwhelmed by the immediate acute problems stemming from the gas leak:

The State and Central government could simply not acknowledge cyanide poisoning [see below] due to its implications: it was beyond the capacity of the administration to care for such a large proportion of physically and mentally ill, most, irreversibly so, and with steadily deteriorating health, and therefore better for it to restrict damage publicly to one or two major symptomatic disturbances, and rule everything else as of no medical significance. (APPEN, p. 4)

Here again there is a conflict of views over the organization and efficacy of such health measures as were being provided in the aftermath of the killing. While all sides agreed that the immediate reaction by medical services and personnel was as good as could be expected, two different versions emerge about the follow-up operation. A member of the US government agency, the Centers for Disease Control in Atlanta, Georgia, Gareth Green, had been invited to Bhopal by the Indian government. He told the *Journal of the American Medical Association* (12 April 1985, p. 2004), 'We felt that, after the initial shock, the medical teams were well organized, with good communications throughout the city.' Green praised the response of the Indian doctors: 'They soon realized that the problem they were dealing with was the effect of the gas on the eyes and lungs. They pretty rapidly formulated standardized protocols to deal with the problem and these were widely implemented.' Green noted these protocols would be of use in long-term studies: 'It will reduce the

Gas Victims' Help and Treatment Centre, run by the Self-Employed Women's Association, or SEWA, which also means 'service' in Hindi. The choice of name was intended to overcome linguistic barriers.

variabilities that might have occurred if a wide variety of treatments had been used' (p. 2003).

It is hard to reconcile this view of an organized, efficient health service with the description provided by radical Indian medical groups working on the ground in Bhopal. The Medico-Friend Circle, a voluntary group of Indian medical personnel, reported a situation totally at odds with that described by Green.

In clinics and hospitals treating victims, no case papers were made out: therefore nothing – neither patients' names, symptoms nor their treatment – was recorded. Senior medical personnel gave no guidelines to doctors treating the victims. No certificates of death or disease were issued. The medical services obtained an incorrect view of the effects of the gas due to the emphasis on clinic and hospital work and due to the failure of the medical profession to organize relief on a community basis. (Medico-Friend Circle Bulletin, *January 1985, p. 5*)

Nor does Green's image of organized efficiency sit easily with Praful Bidwai's report (*Times of India*, 3 December 1985) a year after the killing that proper medical records exist only for fewer than 10,000 victims.

The lack of organization of government clinics and hospitals, with their casual and callous treatment of the gas victims, also served another purpose. Patients waited for hours for treatment, were not listened to, and found the treatment they obtained at the clinics to be not only useless but also demeaning. In response to this, patients lost faith in the government clinics and abandoned them, while flocking to private medical practitioners in droves. This enabled the government to claim that the number of patients who were still ill, defined as those using government hospitals and clinics, was decreasing and thus the health effects of the gas leak were subsiding.

This callous treatment of gas victims by medical staff continued into 1986:

Patients attending specially-set-up clinics are treated with apathy, and sometimes, plain hostility. In a surprise visit last September, the Chief Minister found four doctors absent from a clinic and had to suspend them on the spot. A special committee sent by the Prime Minister Rajiv Gandhi last August to investigate charges of incompetence in relief and rehabilitation programmes concluded that medical facilities were woefully inadequate. (Third World Women's News, Vol. 1, No. 1, p. 16, 1986)

The Bhopal Group for Information and Action reported a boom in private medicine and quackery in Bhopal:

As one private doctor stated very candidly, 'The government doctors prescribe the same medicine as we do, they have much greater facilities; if they happen to improve their services, we will be out of business.' That doctor, like most others operating in the gas-affected bastis *[shanty towns] of Bhopal, does not lose sleep over this possibility, fully aware that such a change in the governmental attitude is extremely improbable. (Bhopal, July 1986, p. 7)*

Conflict of Medical Theories

Green's assertion that the major problems requiring resolution related to the eyes and lungs alone is also open to question. This question is of a part with the assurances that no long-term effects could be expected from the MIC exposure. These assurances, as already mentioned, rested on the claim that MIC would break down into harmless substances on contact with moisture: thus the only effects would be on the lungs and the eyes. This hypothesis was presented not only by the doctors UCC had flown into Bhopal, but also by two members of the US government team from the Centers for Disease Control:

Methyl isocyanate gas breaks down rapidly in the presence of water, becoming a relatively non-toxic substance known as dimethyl urea, points out James Melius, MD, chief of the Hazards and Technical Assistance Branch at the National Institute for Occupational Safety and Health in Cincinnati. (Melius was a member of the Centers for Disease Control group.)

Thus, Melius says, methyl isocyanate is not hazardous for any prolonged period. Indeed, as far as can be determined, the cloud in India dissipated within two hours, Kaplan (also of the Centers for Disease Control) says. (Journal of the American Medical Association, 12 April 1985, p. 2004)

This breakdown hypothesis was later contradicted by the findings of the

Indian Council on Medical Research (ICMR). Yet in Bhopal it held sway, providing the basis for what has been described as the pulmonary thesis.

The Medico-Friend Circle has contended that, after the killing, Bhopal was the scene of a struggle between two medical theories:

> (I) *The 'pulmonary' theory, which believed that in view of the available information about the effects of MIC, only extensive lung damage (leading to diffused pulmonary fibrosis) and direct injury to corneas of the eyes could be expected.*
>
> (II) *The 'enlarged cyanogen pool' theory, which believed that the effect of the released gases on the patients was to increase the cyanogenic pool inside their bodies, leading to chronic cyanide-like poisoning. (Medico-Friend Circle, 1985, p. 1)*

The pulmonary theory was the basis for the chorus of reassurance already described. This theory excluded the possibility of unknown effects arising from exposure to MIC. It ignored the lack of available information on MIC's effects, especially long-term effects. It further ignored the possibility of synergistic effects, that the combined effects of exposure the various gases released would not simply be confined to the known effect of one of the gases, that is, MIC. It ignored the possibility that contaminated or decomposed MIC might have effects different from those described in experiments – extremely limited as they were – with pure MIC. The second theory allows for the possibilities just mentioned and also accepts the effects that the pulmonary theory emphasizes: it does not deny the existence of purely pulmonary effects on health, but acknowledges a mixed pathology.

The struggle between the two theories was more political than scientific. 'Although outwardly the conflict is theoretical, it has little to do with scientific rigour and debate' (Medico-Friend Circle, p. 49). The pulmonary theory obviously answers the need of both state and capital to limit the killing's effects. Thus this theory was strongly advocated by UCC and that section of the Indian medical bureaucracy and profession that had close ties to and sympathy with UCC (Agnew, 1985, p. 19). 'Supporters of [the] pulmonary theory include a dominant faction in Gandhi Medical College, Bhopal, and has strong support in the health department of Madhya Pradesh government. They are adamantly refusing to accept any theory but their own theory' (Medico-Friend Circle, p. 6). This is confirmed by Everest:

In interviews, Drs Misra, Mathur [both of Gandhi Medical College] and Ishwar Dass, the additional chief secretary in charge of co-ordinating health and relief services in Madhya Pradesh, all contended that the effects of the gassing were basically confined only to the lungs and eyes. On 25 December 1984, Dr Dass stated officially, 'No organ of the body has been affected except the eye and lung. There is not a single case of kidney, liver or nervous system damage.' (1985, p. 82)

In contrast, the major proponents of the 'cyanogen pool' theory were the ICMR and various dissident doctors and medical groups operating in Bhopal. The struggle between the theories also reflected a split over the nature of the gases released during the runaway reaction in tank no. 610. The general non-pulmonary theory was shared by all who believed that not merely MIC was released: this included those who argued that hydrogen cyanide was formed during the decomposition of MIC during the runaway reaction, and that this gas was primarily responsible for the deaths and injuries in Bhopal. However, the 'cyanogen pool' theory did not require hydrogen cyanide to have been formed and released during the runaway reaction: it merely argued that the gases released had affected the cyanogen pool within the body, generating an increase in cyanide radicals (free cyanide) inside the body, thus leading to symptoms of chronic cyanide-like poisoning.

Thus, while the 'cyanogen-pool' theory did not necessarily confront UCC over the composition of the released gases, it took issue with the ruling notion, highly beneficial to UCC, that the gases' effects were limited. This theory allowed the Indian state, through the ICMR, to challenge UCC on a major issue, without requiring the state to confront UCC on the issue of the toxic cloud's composition. Moreover, as it was only a theory, it could later be abandoned by the state if it no longer needed to challenge UCC over the leak's health effects – for instance, if a satisfactory deal was reached over compensation.

Both theories had associated problems. The major problem with the pulmonary theory was its inability to account for the actual health situation in Bhopal. Of course, this was not its function: it was more intended to deny elements of the health situation. The Medico-Friend Circle pointed out the theory's inability to explain the range of symptoms reported in their community-based survey. The Medico-Friend Circle survey found only 2.7 percent of people had pulmonary symptoms exclusively, while 62 percent had pulmonary, gastro-intestinal, eye and central nervous system symptoms and 35 percent had symptoms (gastro-intestinal, central nervous system, eye) without any pulmonary symptoms. The exclusively pulmonary theory could in no way account for the very wide range of symptoms reported in the Medico-Friend Circle report: fatigue (86 percent), blurring of vision (77 percent), muscle ache (73 percent), headache (67 percent), flatulence (68 percent), anorexia (66 percent), nausea (58 percent), excessive lacrimation (58 percent), tingling and numbness (54 percent), loss of memory (45 percent), anxiety and depression (43 percent) as well as impotence, shortened menstrual cycle in women, increased dysmenorrhoea and leucorrhoea (Medico-Friend Circle, pp. 40–1).

The Medico-Friend Circle has also criticized the ICMR's 'cyanogen pool' theory on the grounds that the Council has not done adequate research to validate the theory. The double-blind experiment that the Council performed with sodium thiosulphate, the recognized antidote for cyanide poisoning, had shown relief of pulmonary symptoms only:

Going by whatever evidence the Indian Council on Medical Research has published so far, it is not adequate enough to explain the wide range of symptoms in a high proportion of the ambulatory population as revealed in our study. This criticism clearly leads one to suggest that the Indian Council on Medical Research does not have adequate evidence to substantiate the 'cyanogen pool' theory or if it has got it for some unknown reason it has not made public full details five months after the disaster. (Medico-Friend Circle, p. 42)

The Council was to remain ambivalent about the theory it had proposed. By the time it finally issued its summary report in April 1986, a full sixteen months after the disaster, it had become more ambivalent yet: '[The report] says that cyanide was metabolically generated in the body of many of the victims exposed to MIC but that over time the amount decreased' (*Chemical and Engineering News*, 14 April 1986, p. 6).

Limiting the Health Crisis

The state's strategy, then, was to limit the admitted health effects of the killing. Its tactics consisted of denying the existence of symptoms, refusing to recognize new symptoms as gas-related and dismissing other symptoms as normal or otherwise unrelated to the toxic exposure. These tactics were most strongly supported by those among the Madhya Pradesh élite who were most sympathetic to UCC. Thus, the pro-UCC lobby were not satisfied with the pulmonary theory alone. Even the pulmonary effects were played down by attributing pulmonary problems to pre-existing diseases such as tuberculosis.

On the 27th of December [1984] Arjun Singh (chief minister of Madhya Pradesh)[1] *would prominently open a tuberculosis camp in one of the gas-affected bastis (shanty towns). Later in January, he would tell Shahnaz Ankleswaria of the* Statesman *that there was a necessity to establish a 'definite connection' between the recent deaths and the MIC leakage. They could have occurred because of the victims' past poor health. (APPEN, 1985, p. 7)*

This attempt to write off the leak's pulmonary effects as pre-dating the gas leak was strongly supported by Dr Misra, an influential staff member of Gandhi Medical College. As reports of lung ailments and breathlessness continued, Dr Misra told journalists, 'These are mostly previous diseases, which get accentuated in winter. In fact they're known as "winter bronchitis" – people believe it's MIC.' Misra held consistently to this position. Speaking at an ICMR meeting in May 1985, 'Dr Misra . . . was of the opinion that the majority of cases who showed restrictive lung functions were those who had pre-existing lung disease' (APPEN, p. 143). No doubt only the solemnity of the meeting prevented the good doctor from attributing these problems to 'summer bronchitis'.

Such positions were common among doctors in Bhopal. The Medico-Friend Circle reported in April 1985 that fatigue among gas victims was being dismissed as 'laziness' and 'wanting to make the best use of the aid pouring in', while respiratory symptoms due to lung damage (fibrosis) were being incorrectly diagnosed as tuberculosis and psychological effects were being seen as 'compensation malingering'. That this viewpoint was common among doctors dealing with the gas victims is shown by an interview with an unidentified Indian doctor in *Comhlamh News:*

As soon as people realized, however, that compensation was available, the hospitals which had to cope with the emergency were inundated with thousands of residents demanding treatment and a certificate to state that their health had been adversely affected by the gas. 'The majority of these people', claimed the doctor, 'were impostors, but it was impossible to remove them from the hospitals. The serious cases, those people who needed immediate treatment, had to wait before the handful of medical staff could attend to them . . .' The promise of compensation, he claimed, has also given rise to many exaggerated and unconfirmed reports of thousands of victims suffering long-term effects of the gas. In the few days after the leakage the eyes of many people were red and irritated. This became exaggerated into claims that the gas had caused irreparable blindness, a cry eagerly taken up by newsmen, anxious for a more dramatic story. Bronchial patients, he recalled, who had lung problems for many years, would cough and wheeze every time a reporter passed. The figure of 100,000 injured, which appeared in some papers, was, in his belief, a wild exaggeration. (Autumn 1985, p. 27)

Thus the illiterate population supposedly conspired together to construct their diseases as a plan to milk money from the unfortunate US multinational. This exceptional conspiracy theory comes up against one major problem. The people of Bhopal continued to suffer from their diseases, even after the immediate relief operation was over, even after the hope of receiving money from America receded from view. The majority of those affected by the leakage were day labourers, who were paid by their daily work. They had no investments or property to live on while they carried on their conspiracy. They would work if they could, as this was their only access to money. Many of them correctly expected that they would not get a rupee. It is indeed a powerful conspiracy that enables these people to ignore immediate problems such as obtaining food and money, living on thin air and the prospect of manna from America, which would arrive, like the cargo ships of South Pacific cultists, at some future time.

Still, this blaming the victims, or blaming the victim's environment, or blaming anything but the gas leak, was to be a constant refrain of the medical establishment. It reached its nadir in Misra's statement, to *Business*

Week in November 1985, that a large number of Bhopal residents are not physically sick but are suffering from neurotic reactions: 'They think they are weak and can't concentrate, but when we give them exercise tests, they do fine.' Yet even Misra had to admit some 60,000 people had a lowered capacity for work due to the gas leak. His psychological argument shaded into one that people are faking illness to obtain compensation. Some doctors are reported to have refused to issue death certificates to gas victims' relatives, alleging that they only wanted compensation. The Bombay magazine *Sunday* alleged criminal negligence on the part of Bhopal hospitals, citing cases of women, asking for their children's cause of death to be entered on death certificates, who had their death papers confiscated so that they were unable to claim compensation.

Doctors, with much of the middle class, were able to maintain their narrow views because of their class and caste prejudices, as well as their practice being confined to clinic and hospital work.

At the 100th Day ceremonies in Bhopal government officials reported that the medical situation was slowly returning to normal. Affluent Bhopal citizens who never go into the slums may have believed it . . . Most of the wealthier residents whose neighbourhoods were inundated by the gas, if they were not killed in the first wave, were able to get out of the city by car before being overwhelmed. These are the lucky ones whose symptoms began to disappear in the following weeks, because they had immediate access to vehicular transportation and were not forced to run in their efforts to escape. (Dakin, 1985, p. 4)

Struggle Over Reproductive Effects

One major health effect, upon reproduction, became an area of struggle. The issue was raised in the early days after the killing. As early as 3 January 1985, the Morcha was demanding that 'Economic assistance at par with what is given to the families of the dead should be provided to gas-affected women who suffer from abortions, or give birth to stillborns or deformed offspring.' While the medical bureaucracy did consider the question of reproductive effects, it was only in terms of the health of unborn children: the unfortunate mothers' health was not considered. Nor did the medical establishment consider that the fear of teratogenic effects required them to warn women of the possibility. This would only create 'confusion and panic' and demands for abortion. The medical establishment argued that there was no need to warn of this possibility, in the absence of proof that such effects would occur. This ignored the cases of women who suffered abortions immediately following the gas leak, which should have been interpreted as a warning. The medical establishment did not even institute screening procedures to check on this. Mira Sadgopal reports:

A number of [women] said: 'We wondered whether there was a risk. We tried to get ourselves examined but the doctors just threw us out with some tablets and said everything was all right.' The irony is that these doctors, or others who are part of the same system, are doing research to measure expected foetal defects. I should say they are expecting defects. On the one hand, they say there is no proof of damage, and on the other, they are getting ready to measure it for their research papers.

This position did not go unchallenged. In mid-February the Medico-Friend Circle called for the potential dangers to the foetus to be publicized. They called for adequate facilities for ultrasonography and amniocentesis to be provided.[2] Beginning on 8 March, International Women's Day, the Morcha took a travelling exhibition through the shanty towns, with the intent of providing information to women, particularly on the question of abortion and the need to use contraception until the gas's reproductive effects were known. These actions succeeded in raising the issue. As Padma Prakash later noted, 'the concerted attempt of local activist groups and women's groups to provide relevant information in the *bastis* (shanty towns) was only successful in creating a better understanding about the problem *but could not resolve it*. No infrastructure for alternative abortion or ultrasonography facilities could be established' (*Economic and Political Weekly*, 14 December 1985, p. 2196). Those women who did take the initiative to seek abortions found obstacles put in their path, despite India's previous callous history regarding its subjects' fertility and the official availability of abortion on request as part of India's birth-control policy.

It's worth noting, here, the gap between the private opinions of doctors and those they expressed publicly. Mira Sadgopal related in an interview:

These very same doctors, government doctors, in particular in government service, in public they were not saying, but in private when you talk to them said that they did feel there was a risk and when I specifically asked them what they would do if they were in the position of one of the women, they said that they would not even go in for amniocentesis or ultrasonography, because what would that tell you? It wouldn't tell you so many things, mental retardation or whatever, we would go straight for MIP (medical interruption of pregnancy) because it's a relatively safe thing, and temporary psychological trauma weighted against the possibility of raising a child which might be deformed, we just don't know. (APPEN, 1985, p. 46)

Despite these private doubts, the doctors held the official line in public. A study by two women doctors associated with the Medico-Friend Circle found an epidemic of gynaecological effects among gas-exposed women, but these were dismissed out of hand by the medical establishment. Drs Bang and Sadgopal wrote:

We had an opportunity to talk with three gynaecologists in the field clinics and one professor of Ob-Gyn in the hospital, all of whom stated that there are no gynaecological problems attributable to the disaster. They explained the large number of women complaining of gynaecological symptoms as 'usual', 'psychological' or 'fake' and the gynaecological diseases in these women as 'usual', 'tuberculous' or 'due to poverty and poor hygiene', refusing to accept any special situation. (Bang and Sadgopal, 1985, p. 7)

The medical establishment continued to deny reproductive effects. In January a spokesperson for the Sultania Women's Hospital denied there was any increase in abortions attributable to the gas exposure. Such reporting of reproductive effects as did occur showed a class bias:

It was expected that many deformed full-term babies, first trimester foetuses on 3 December, would be born starting in June 1985. An uncounted number of horrifying births have already taken place among poor women, attended only by midwives, but the only one which I saw in the papers was born in a hospital in Bilaspur, to a woman who had fled Bhopal the day of the disaster. She was the wife of a Bharat Heavy Electricals Ltd executive in Bhopal. (Dakin, 1985, p. 7)

When the increased abortion rates could no longer be denied, it was argued that this increase was not directly related to the gas, but was in fact caused by increased stress which was associated with all disasters, and it would end when the immediate stress associated with the gas leak ended. This was untrue as well: the Medico-Friend Circle study of the gas's reproductive effects showed the effect of the gas leak on pregnancies continued after the acute exposure period; it continued to affect pregnant women even ten months after the gas leak. Nevertheless, those wishing to exonerate the gas had still one last argument: the gas's reproductive effect was handily complicated by the mishmash of powerful drugs dispensed to the victims for symptomatic relief. Any rigorous scientific study would have to control for this exposure – no easy task. Perhaps this explains why 'The Indian Council on Medical Research was forced to scrap their pregnancy outcome study thrice as both the design and the date were faulty' (*Medico-Friend Circle Bulletin*, October 1985, p. 4).

Problematic Research

Tight control of information was essential for the state to maintain control of the definition of the limits of the killing. Thus it was necessary for the state to obstruct and repress independent research. Padma Prakash reports: 'The "people's scientists" have had serious problems in getting the samples tested – laboratories have not been willing to release reports and some of the sampling techniques needed to be innovated, some need sophisticated equipment only available in government laboratories.' Similarly, women's groups planned an independent survey of the killing's teratogenic effects for June, involving forty women researchers from outside Bhopal, but it took place only in September; the delay was caused by government repression.

Scientists were also discouraged from working with groups active in Bhopal. Dr Sadgopal told APPEN (1985, p. 49) of a scientist who was working with the Morcha to produce a simple test for thiocyanate detection: 'After some time he said he would be unable to go on. His own institution did not appreciate his meeting people like us.'

When it became obvious in February 1985 that an immediate settlement with UCC was unlikely, the Indian state began its own research programme. The head of the ICMR denied a statement previously attributed to him that 'there is no reason to believe that there will be any long-term effects' and announced a major programme of research. Investigation of the actual accident resumed in February 1985. According to the Varadarajan report, the first samples of residue were taken from tank 610 on 20 December 1984; presumably this was when UCC obtained its samples. Additional samples were taken in February 1985. In March 1985 the tank was excavated. In April 1985 further sampling of residue took place and the tank itself was examined. UCC responded to this research programme in March by announcing its own series of animal tests intended, no doubt, to bolster its own position.

Writing on the Indian scientific community's response to Bhopal, Padma Prakash said they 'reduced a social catastrophe to a scientific/clinical/medical problem'. Even if they had been successful in this limited project, it would have been of some use. Yet the majority of the ICMR's projects have been rightly criticized as too academic and of little use in tackling even the immediate clinical problems, without considering the psychological or social problems.

By May 1985, the Council had already funded twenty research programmes at a cost of 1,500,000 rupees. All but two of these studies were to last two years. Their main aim was to 'gain knowledge on the delayed and long-term effects of exposure which would be of value in treatment and in a better outcome eventually'. The time-scale of the studies meant that any benefits derived from them would be late in arriving.

The two short-term studies were hardly more useful. One, which was a general follow-up, confirmed other studies of the gas leak's effects: 74 percent suffered from coughs, 34 percent from breathlessness and 36 percent from burning eyes. But, while the study's protocol stated that treatment would be provided, no treatment was mentioned in the March summary report.

More intriguing is the study of ocular changes, which concludes that there are no cases of complete blindness and that the majority of cases have completely recovered with treatment. This is surprising, because the King Edward Memorial Hospital/Voluntary Health Association of India study reports a high percentage of visual problems in children. At the end of March when the Medico-Friend Circle team visited Bhopal to conduct their study survey, eye disorders seemed widespread. (Prakash, 1985)

Prakash also notes that the ICMR preliminary report on teratogenic effects omits the high reported still birth rate, for which a baseline rate exists for comparison purposes.

The Medico-Friend Circle also strongly criticized ICMR projects, noting they were focused on the seriously ill hospitalized patients and specifically on pulmonary effects. The more diffuse and generalized effects in the non-hospitalized population were ignored. The Medico-Friend Circle also strongly condemned the Council for not undertaking research to validate its 'cyanogen pool' theory.

The serious gaps in the very fabric of research efforts have direct and vital connection not only with the urgent issue of relief from suffering, damages and compensation to the victims of the poison gas, but also with the issue of fixing the responsibilities on all those who have perpetuated the suffering of thousands of people of Bhopal. (Medico-Friend Circle, 1985, p. 2)

ICMR researchers on the ground were unable to gather information on unexpected effects, which were not anticipated and included in their study protocol. Mira Sadgopal relates two incidents of this: 'We then told the ICMR male recruit to ask about this aspect [male impotency] in the course of the survey. He was in a quandary because the proforma did not have any space to record this information so the computer could not take it in.' Also, 'we reported it [higher pulse rate] to the ICMR. Again, they said their computer was not programmed to take in this information.' Some proposed research studies were obviously biased. Gandhi Medical College, where Dr Misra is head of medicine, proposed the following study:

One of the proposals aims at studying pregnant mothers exposed to MIC to see its effects on the foetus and on the newborns. The sample considered in this study is clearly skewed in favour of middle-class colonies [estates], while poorer localities have been left out of the study. Thus if the effects of MIC on pregnancy have any relation to the mothers' bodyweight or nutritional levels, then the results of this study are unlikely to capture them. (Business India, 25 February 1985)

This is only to be expected from Misra, as he gave the following details of animal experiments he claimed he had undertaken to study the gas's teratogenic effects: 'We caught rats from the gas-exposed areas and mated the female rats with healthy males. The progeny was found to be perfectly healthy, even though the mothers had lesions on their lungs.' (*Business India*, 25 February 1985).

Other research was hampered by the conditions in which it took place. Following up sodium thiosulphate therapy in the thirty-bed Deputy Inspector-General hospital provided for the ICMR by the Madhya Pradesh health authorities was made difficult by the actual conditions in the hospital. The *Indian Express* reported (27 March 1985), 'Till a few days ago this hospital was functioning without a refrigerator in this city's blazing heat. The machine which analyses blood gases is exceedingly sensitive to any rise in temperature because it contains a computer, yet the administration has stopped the air conditioners. This can render the test results completely unreliable.' Other studies were reported to have taken place without the informed consent of the patients. A King Edward Memorial Hospital study, in which a series of blood tests and lung biopsies were performed on fifty exposed railway workers, allegedly took place without the patients' permission. The Medico-Friend Circle (1985, p. 50) also attacked the lack of informed consent among those being treated experimentally by the medical establishment in Bhopal.

This lack of concern on the part of the medical and scientific bureaucracy also showed in their lack of interest in providing information to the gas victims. Neither the ICMR nor the state showed any interest or inclination to give information to the gas victims, though basic health information and simple breathing exercises would have been of immediate practical use. The ICMR told 'individual researchers not to publicize their findings prematurely and individually so as to avoid confusion and panic in the public mind through conflicting statements'. This attitude extended towards voluntary health groups as well. According to the Medico-Friend Circle (*Bulletin*, September 1986, p. 1), 'it was almost next to impossible to get any information from the medical establishment in Bhopal. This conspiracy of secrecy was extended to such ridiculous lengths that even such innocuous information as the ICMR numbering and the maps of the *bastis* was treated as classified documents.'

This policy of restricting information was to continue. The chairperson of an ICMR meeting on 3/4 May 1985 called for the early publication of the findings of research projects on exposure to the toxic gases. The chairperson suggested accelerated yet peer-reviewed publication in the *Indian Journal of Medical Research*. While one member at the meeting cautioned the meeting about the possible legal implications, the consensus was that the findings should be published as early as possible. Researchers were asked to have articles produced by July or August 1985. The ICMR finally published a summary report of Indian scientists' findings in April 1986. On the same theme, it was reported in the *Economic and Political Weekly* that:

In March 1985 an ultra-modern, fully computerized, sophisticated instrument complex which can carry out thirteen tests from a range of twenty-five parameters simultaneously at a phenomenal speed, releasing every fifteen seconds precise and accurate data which can be stored indefinitely and retrieved at a moment's notice, was loaned to Hamidia Hospital . . . the paucity of information released by the government indicates that either the machine is not being properly utilized or the information about the large number of patients dealt with by the machine [is] treated as classified. (14 December 1985, p. 2192)

This official silence did not go unchallenged. Beginning on 8 March 1985, the Morcha took a travelling exhibition through the shanty towns to provide information on the gas leak's effects. The exhibition was aimed originally at women only and it emphasized the need for contraception and the abortion option. The desire for information among the victims was so great that part of the exhibition had to be opened to men as well. The medical establishment's desire to keep information from the people was based on an élitist vision of medical knowledge, which it believed would not be understood by the affected people. It was part of the state's general strategy to limit and control the health crisis: the results of 'confusion and panic' from the provision of information would undoubtedly be increased demands for treatment and other facilities.

The failure to undertake the most basic scientific research that followed from the killing had major results. Praful Bidwai stressed that one reason for the desultory medical relief in Bhopal *is the failure of the Indian scientific and medical bureaucracies to analyse the composition of the material released from the Carbide plant,³ to study the effects of numerous toxins on the body, and to devise appropriate lines of treatment to cure the victims – a failure that is related to the weakness of the Indian technological infrastructure and to the close links that Carbide managed to establish within that apparatus.* (Inside Asia, *November 1985, p. 42)*

Prakash (1985) observed: 'The fact that no definite research-based report on water and soil has been forthcoming even six months after the disaster is disturbing.' To this should be added the Medico-Friend Circle's indictment of research sponsored by the ICMR: 'It is a sad commentary on these research efforts that five months after the disaster with a mass of population continuing to complain of serious symptoms, no comprehensive picture of morbidity pattern in the community is put together either by the ICMR or the medical establishment of Bhopal' (Medico-Friend Circle, 1985, p. 47). The general failure by India's scientific community leads Prakash to conclude: 'Never has the "world's third largest population of scientists" seemed more like a deadweight as now, in the wake of Bhopal.'

Similar problems beset the research on MIC itself. While one of the killing's undoubted results was a major research effort on MIC's toxicology, this seemed a case of shutting the stable door after the horse had bolted. Furthermore, in the way the research was concentrated on MIC *only*, the research cannot be expected to be of much more than academic interest. Such academic research has already been misused, however. A study undertaken by B. Nemery and others at Britain's Medical Research Council Toxicology Unit was reported in the general and occupational health media as disproving Indian claims that the cyanide antidote, sodium thiosulphate, was effective in treating the killing's survivors. Nemery and co-authors concluded: 'This study suggests that $Na_2S_2O_3$ (sodium thiosulphate) does not protect rats from the acute and subacute effects of MIC and that cyanide intoxication is not involved in the clinical syndrome seen after exposure to *pure* MIC' (*Lancet*, 30 November 1985, p. 1246, emphasis added). It doesn't occur to these researchers to note that pure MIC is a different kettle of fish from what hit the victims of Bhopal. If anything at all is obvious from the reports from

India, it is that the gas victims were not exposed to pure MIC. Even ignoring the question of cyanide contamination, the gases were the result of a runaway reaction involving industrial MIC, contaminated with a higher than normal level of chloroform.

While Nemery and others made no protest over the use of their research to discredit sodium thiosulphate therapy, other toxicologists were more circumspect in pointing out the limitations of their research. Toxicologist Dr John Bucher, who led the research on MIC for the US National Institute of Environmental Health Science, pointed out that their research dealt only with pure MIC, while 'in the Bhopal accident as much as one-third of the materials released may have been created by the explosion [sic]. Effects of these materials were not researched' (*Health & Safety at Work*, August 1986, p. 7).

The problem here is one which is common to much toxicological research, which tests one chemical in isolation. Yet exposure, normally of workers but in this case also of the general community, is usually to a mixture of chemicals. Thus the research ignores the synergistic effects of exposure to more than one chemical, the way that chemicals' effects can multiply each other and interact with each other. Commenting on studies performed by the Indian Toxicology Research Centre,

Indian toxicologist C. R. Krishna Murthi, a former Indian Toxicology Research Centre director, warns that all studies so far have been based on a single chemical gas, MIC. More studies should be conducted, he says, to examine the effects of MIC as part of a mixture of gases, because the plant leak may have involved other gases besides MIC. (Chemical Week, 4 December 1985, p. 12)

William E. Brown of Carnegie-Mellon University also stressed that the synergistic aspect is being ignored in post-Bhopal research:

One of the things that there's less emphasis on, in the research effort we know about, is the toxic potential of methylamine (formed from MIC in a moist environment) and methyl isocyanate. If you have these two together, you can get an adduct of the two. If there was something like cyanide in the air and in combination with the other two, what would be the effects? It's *opening up a whole new area of research. (Chemical and Engineering News, 14 October 1985, p. 43)*

To these problems must be added doubts over the applicability of animal experiments to humans, apart from the appalling arrogance involved in such cruel use of our fellow animals. Finally, this type of research abstracts the effects of chemical exposure from their social context. In the case of Bhopal, lack of nutritious food, lack of work and income, frustration, anxiety, depression and many other factors impinge on the health of the gas victims. These factors get lost in abstracted, pure research on one chemical's effects on experimental animals, just as the concentration on one chemical is abstracted from the reality of mixed exposure to chemicals that occurred in Bhopal.

Politicization of Treatment

In the period immediately following the killing there was major confusion over what chemical or chemicals had leaked from the Bhopal plant. Since UCC were denying MIC was toxic, a theory grew in the first few days that phosgene was the gas that leaked. This theory was supported by reported statements of such foreign experts as Dr Stuart Luxon, Health and Safety Director of the Royal Society of Chemistry, who stated, 'All the symptoms and all the effects point directly to phosgene poisoning.' Sir Frederick Warner, who had headed the inquiry into the Flixborough chemical-plant disaster in Britain, said descriptions of a heavy poison gas cloud fit phosgene rather than MIC, which is twice as light. This theory encouraged treatment of symptoms only.

The phosgene theory, however, as some doctors and other investigators have pointed out, had larger implications. It successfully diverted attention from the toxicity of MIC, a base chemical on which Union Carbide as a company has built its chain of pesticide products. The Bhopal plant itself was to launch from December a range of new products, including a formulation similar to Baygon, the household pests spray, all of which were based on MIC. Many chemical corporations are raised on the exploitation of a single base chemical and the toxicity of MIC threatens Union Carbide's empire world-wide. What the phosgene hypothesis immediately achieved was to confuse doctors to the extent that no attempt was made to find or administer or to even consider the question of an antidote seriously. While the experts debated, people died like flies. (Illustrated Weekly of India, *31 December 1984*)

Within a few days, phosgene was replaced by MIC as the responsible gas, but the general belief, encouraged by UCC doctors and other health experts, was that MIC had no long-term consequences. This theory ran into practical difficulties when Dr Heeresh Chandra, a pathologist at Hamidia Hospital, suggested that, on the evidence of the autopsies he had conducted, the victims might have died from cyanide poisoning. Dr Chandra suggested that victims should be treated with sodium thiosulphate, a recognized treatment for cyanide poisoning.[4] Investigators from the Central Bureau of Investigation were also reported to have found cyanide in blood samples from gas victims.

From December 4, a forensic expert attached to the Central Bureau of Investigation in Delhi conducted several tests on blood samples drawn from the victims over a week. His results showed either the presence of MIC directly, or of cyanide and of monomethylamine – the products into which the gas had been broken down in the body. (Illustrated Weekly of India, *30 December 1984*)

Dr Chandra gained some local support. But, as the *Times of India* (11 December 1984) reported: 'Sodium thiosulphate was recommended right in the beginning by some Hamidia doctors, but they were dissuaded by Union Carbide's chief medical expert, Dr Avashia, who told them it was neither necessary nor advisable.'

Dr Chandra received support from foreign experts as well. A telex from the US Centers for Disease Controls received in Bhopal on 11 December 1984 devoted half its contents to the treatment of cyanide poisoning (APPEN, 1985, p. 7). By 8 December 1984, Dr Max Daunderer, a clinical toxicologist with the Munich Institute of Toxicology, West Germany, who had been invited to Bhopal by the Indian government, agreed with Dr Chandra, having found cyanide levels of 2 ppm in patients' blood samples and patients responding to sodium thiosulphate. Although Dr Daunderer had brought 50,000 doses of sodium thiosulphate with him, he was not allowed to administer them. Dr Bhandari, medical superintendent at

Hamidia Hospital and Dr Misra, vice-dean of Gandhi Medical College, demanded analytical proof of cyanide poisoning before they would allow sodium thiosulphate to be used. When Dr Chandra pointed out such an analysis might prove inconclusive, he was told by Dr Misra, 'You are a doctor of the dead, so do not interfere with the living.'

In a situation where overmedication for symptomatic treatment by dangerous drugs was in full swing, the local medical establishment curbed the use of a reportedly harmless drug, demanding an obsessionally ritualistic scientific proof before they would accept its use. This is comparable to the reaction of Madhya Pradesh chief minister Arjun Singh on the morning of the killing: he was not prepared to do anything until he had the 'precise technical details', whatever they were. In effect Mr Singh was awaiting the delivery of the body count, rather than intervening and attempting to reduce it. In both cases, the need for a fetishized version of technical proof was advanced as a reason for inaction. In Dr Misra's case, even the science was wrong.

In treating suspected cyanide poisoning, clinical evidence is normally considered sufficient basis on which to begin treatment; it is not necessary to await a laboratory report. Yet that is what the medical establishment in Bhopal was demanding before it would allow treatment with sodium thiosulphate.

On his own initiative, Dr Daunderer administered sodium thiosulphate to two of the worst-affected victims. One died and the other showed a spectacular recovery. The patient who died was later shown to have such extensive lung damage as to have been unable to respond to any therapy. In this case, however, the anti-sodium thiosulphate lobby did not wait for such scientific niceties as laboratory tests – and possibly autopsies. Rumours were spread that 'the German doctor has killed a patient using sodium thiosulphate'. *Business India* reported that Dr Daunderer was 'bundled out of Bhopal'.

Dr M. Nagu, Madhya Pradesh government director of health services, sent out a circular stating, 'under no circumstances shall sodium thiosulphate be given unless it is correctly and conclusively proved in the laboratory that it is cyanide poisoning'. Dr Nagu said the ICMR had phoned him from New Delhi to request such a circular be issued, but the ICMR denied issuing any such instruction. The day after Dr Nagu's circular was issued, an ICMR circular for Bhopal doctors included a pamphlet on sodium thiosulphate use, thereby implicitly approving its use. Yet his bureaucracy delayed for three weeks before releasing this circular (*New Scientist*, 28 November 1985).

Business India later suggested that the behaviour of Drs Nagu, Misra and Bhandari was due to their 'close associat(ion) with UCIL' and their membership of the company's panel of doctors. That the doctors involved were highly sympathetic to UCC was widely reported. *Sunday* (7–13 April 1985) reported that Dr Nagu was under investigation by the Central Bureau of Investigation for his role in concealing information relating to UCC. It pointed out that Dr Nagu's brother held the security contract for UCIL in Bhopal. Dakin (1985, p. 9) reports an official in the Madhya Pradesh Health Ministry as asserting that Dr Misra 'is still on the payroll as a consultant to Union Carbide'. It is of some relevance here that the majority of studies Dr Misra used to refute Dr Chandra's claims of cyanide poisoning were written by none other than Dr Bryan Ballantyne, UCC director of applied toxicology and chemical-warfare expert.

The sodium thiosulphate camp had originally received support from an unexpected source. UCC's medical director, Dr Avashia, sent a cable from the USA to Bhopal a few days after the killing advising sodium thiosulphate treatment 'if cyanide poisoning is suspected'. But ten days after the killing at a press conference in Bhopal, Dr Avashia's line had changed: there could be no cyanide poisoning and sodium thiosulphate is no antidote to MIC exposure. Avashia said it was due to his mishearing the chemical involved on the radio: he had supposedly confused cyanide with cyanate. When asked why he had suggested sodium thiosulphate, Avashia said he could not know what gases were used in the plant. When pressed on the fact that his cable had been titled 'Treatment of MIC pulmonary complications', Avashia had no answer. On this incident, Dakin (p. 8) reports: 'Dr Avashia turned up in Bhopal, surrounded by UCC officials, to give a press conference at which he said he was mistaken, there was no chance of cyanide poisoning. Bhopal observers say he seemed to be under some extreme pressure to change his story.'

The Avashia incident had yet another ironic twist to it. *Nai Duniya* (Indore) reported (10 December 1984) that 'a couple of days after the incident [*sic*] American experts had prescribed a treatment but the relevant medicines are not available in India. Later it was learnt that these experts of Union Carbide shall bring the medicines for about 500 patients, but no one knows whether they have reached [Bhopal] or not.' While Avashia publicly disavowed sodium thiosulphate treatment, '[Hamidia] hospital sources say that 500 injections of the antidote sodium thiosulphate had arrived from the USA, but that they were distributed amongst the élite and ministers' (*Nai Duniya*, 15 December 1984).

The question of possible cyanide poisoning and the efficacy of sodium thiosulphate treatment continued as a major area of conflict. UCC has consistently maintained its line, following Avashia's press conference, that no cyanide poisoning could have occurred and that no hydrogen cyanide was released in the leak from the Bhopal plant. This claim was met with some disbelief in India. The *Indian Express* (25 January 1985) reported that the US Occupational Safety and Health Agency guidelines for occupational exposure to MIC warn, under the heading of reactivity, 'hazardous decomposition products: toxic gases and vapours (such as hydrogen cyanide, oxides of nitrogen and carbon monoxide) may be released'. The *Indian Express* commented that it is deliberately misleading to ignore this warning and to claim that MIC normally degrades into harmless substances. The London *Times* reported (17 January 1985, p. 12): 'According to evidence from an American scientist given much currency by the factory owners, Union Carbide, methyl isocyanate is distinct from cyanide and is not absorbed into the blood. It is admitted, however, that this has been experimentally tested only outside the body.'

In June, Avashia reiterated that there could be no cyanide poisoning, though he now admitted that there could have been hydrogen cyanide among the gases that leaked. Later again, Avashia told *Chemical and Engineering News* (22 July 1985, p. 6) that reports of 'chronic' cyanide poisoning did not make sense: 'It isn't conceivable that any significant amount of cyanide could have been produced to cause any widespread problem. Medically we had no cyanide poisoning. *But what bothers me most is that people are still being treated with sodium thiosulphate* and there's no justification for that. MIC already does enough damage. Why look for a different scapegoat?' (emphasis added). It seems strange that Dr Avashia was so concerned over the use of a non-toxic drug, when he had raised no objections to the use of dangerous drugs such as cortico-steroids in the treatment of the gas victims.

In the meantime, others supported the UCC line. Dr Gareth Green of the Department of Environmental Health Sciences at Johns Hopkins University, Baltimore, Maryland, in the USA told the *New York Times* (10 April 1985) 'there is no known way of converting MIC into cyanide': thus he claimed there was no scientific rationale for using sodium thiosulphate. This was another example of dismissing a question without examining it. Harold Teague of Pembroke State University, North Carolina, has produced a hypothesis on how MIC could be metabolized in the body to cyanide (Teague, 1985). In reply, Dr Misra told the *New York Times* (10 April 1985), 'There is something called science and there is something called science fiction. This is science fiction.'

In Bhopal, Dr Misra continued to object vehemently to the use of sodium thiosulphate. During a meeting on pulmonary medicine in March 1985, Dr Misra claimed sodium thiosulphate should not be used, as it caused adverse reactions, according to a study he had performed. The study in question was one Dr Misra had presented to the Indian Council on Medical Research in February: in fact it said that only two out of 200 patients given sodium thiosulphate by Dr Misra had shown adverse effects. Dr Misra's conduct during the sodium thiosulphate controversy led to allegations that he had falsified the data.

The Medico-Legal Institute of Gandhi Medical College asked Professor Misra's department to conduct studies concerning thiocyanate excretion in MIC and non-MIC patients. When the final data were sent back to the Institute, the results turned out to be mystifying. In many cases, thiocyanate excretions had increased without NATS (sodium thiosulphate) injections. When the Institute staff called up the original requisition forms, they discovered major discrepancies between the requisition forms and the final list. As it turned out, in every case where thiocyanate excretions had increased, the people had been given NATS injections, but the column recording the number of injections indicated 'zero'. It was also discovered that many non-MIC victims, belonging to the 'control' groups, had been given NATS injections. Such fraudulent registration of data would have disqualified a researcher for life in any respectable academic community.

However, the same doctor, who has called the Indian Council on Medical Research a 'national fraud', was now directing an Indian Council on Medical Research research project on thiocyanate excretion, the results of which the entire medical community knew beforehand. (APPEN, 1985, pp. 9–10)

The reasons for denying the possibility of cyanide poisoning are obvious. Anand Grover of the Bombay Lawyers' Collective charged: 'There was a deliberately engineered controversy by Union Carbide so as to ensure that cyanide would be out of the picture entirely so that any kind of liability, which would be criminal, would not arise at all' (Jackson et al., 1985, p. 6). Dr Mira Sadgopal explained to the *Sunday Times*:

It matters in the litigation for compensation. Union Carbide's lawyers will obviously try to reduce liability as much as possible. An important part of their strategy will be to demonstrate that the industrial slums of India are endemic with tuberculosis and that doctors can't differentiate between tuberculosis damage and gas damage. But cyanide toxicity can't be explained away in terms of an Indian epidemic. It will unambiguously establish the relationship between the gas leak and the thousands who suffered. (1 December 1985)

The criminal aspect is also important. Officials of Film Recovery Systems in the USA were recently convicted of murder after the death of a worker poisoned by cyanide: 'The cyanide issue may make an out-of-court settlement legally impossible if careless handling of a cyanide-related substance is considered to be murder' (Dakin, 1985, p. 11). According to Bombay Lawyers' Collective member Anand Grover, 'If you have knowledge that a person can die with the type of system that you're working with, in India, and in most common law countries, it is culpable homicide not amounting to murder. So it is a very serious criminal act' (Jackson et al., 1985, p. 6).

People's Health Clinic Initiative

The struggle concerning medical treatment of the gas victims continued to centre on sodium thiosulphate. In February 1985 the ICMR had recommended the use of sodium thiosulphate for three-quarters of the exposed population. It stated:

The toxicological studies carried out so far . . . have clearly shown that, at least in the survivors, there is evidence of chronic cyanide poisoning operating as a result of either inhalation of hydrocyanic acid or more probably subsequent generation of cyanide radical from the cyanogen pool in gas-affected victims. The changes in urinary thiocyanate and in blood gas before and after this thiosulphate treatment substantiated the above findings.

Yet the Madhya Pradesh government made no attempt to make treatment available. On 17 and 18 February 1985, the Morcha organized an all-India convention in Bhopal on 'People's Right to Know and People's Right to a Safe Environment'. The convention called for mass detoxification, provision of abortion, contraception and information to the gas-affected people. Still the Madhya Pradesh government did not act. Organizations working with the gas victims only realized in March 1985 the importance of sodium thiosulphate and began to demand its administration. On 24 March 1985, the Delhi women's group Saheli organized a demonstration by 150 women at the Deputy Inspector-General Hospital in Bhopal, demanding proper treatment, including administration of sodium thiosulphate.

Hitavada (25 March 1985) reported that Dr Nagu, Madhya Pradesh director of health services, 'did not give an assurance about sodium thiosulphate treatment, saying he had not received any communication to that effect from [the] Indian Council on Medical Research. He said he would make available sodium thiosulphate ampoules as soon as he received instructions from the Indian Council on Medical Research.' Only at the end of May did the government make sodium thiosulphate available to non-governmental organizations operating in Bhopal, such as the Red Cross, the Indian Medical Association and the Jana Swaasthya Kendra. Even then, sodium thiosulphate was provided only for 'selected patients' rather than as a 'mass detoxification' programme.

The Jana Swaasthya Kendra (People's Health Clinic) was a joint initiative of the two activist groups already mentioned, the Morcha and the Nagrik Rahat aur Punarwas Committee. It was also supported by the Union Carbide Karmachari Sangh, the Bhopal union not affiliated to Rajiv

Gandhi's party, and the Trade Union Relief Fund. The Jana Swaasthya Kendra (JSK) was opened on the tennis court inside the UCIL plant. It was seen as an example of what the government should be doing, as the voluntary groups had realized that the scale of the disaster was such that only the government had enough resources to tackle the problem. In contrast to government clinics, the JSK provided information to people and listened to people's problems and experiences. By the end of three weeks, 1,000 people had been injected with sodium thiosulphate. The JSK and its constituent organizations also emphasized the need for gas-affected people to organize to demand their rights from the government.

That the JSK clinic was a successful initiative was shown by the resulting government repression. In a midnight raid on 24 June, the police arrested doctors and paramedics of the JSK and confiscated JSK equipment and records. The doctors and activists were 'arrested on blatantly false and trumped-up charges like 307 [attempt to murder], rioting and inciting people to riot, etc.' Eight activists were kept in Bhopal central gaol until as late as 6 July. That harassment was the main aim of this police operation was shown by the old tactic of continuing to fabricate new charges to keep activists in gaol: 'When bails were obtained by sympathizers for the existing charges, the bail-amount was increased by the new ones, which were equally false. This was with a clear intention of harassing the people . . . Some *basti* [slum] people, including women with infants, were in gaol more than ten days after the arrests' (Bombay *Daily*, 21 July 1985).

This repression was effective in diverting energy from organizational and relief work. 'All of our energies were absorbed in legal and civil liberty type work for the past five or six weeks. It is only after having spent a great deal of energy and money of the country that we are able to function again and look beyond our immediate problems,' the Morcha wrote on 6 August 1985, explaining why the most recent meeting of the National Campaign Committee had had to be postponed.

There were two responses to this harassment. First, on 25 June a large demonstration of 5,000 people was organized by the Morcha; police responded to it with clubs and arrested some 400 demonstrators. Second, on the legal front, the Bombay Lawyers' Collective took a case to the Indian Supreme Court, which directed the Madhya Pradesh government to make sodium thiosulphate available to the JSK, to appoint a commissioner to oversee medical relief measures and to set up a 'time-bound' detoxification programme.

When the doctors returned to the JSK premises to reopen the clinic, they found it was locked. The Central Bureau of Investigation claimed UCIL had locked it. UCIL claimed the Central Bureau of Investigation had locked it. The doctors were afraid to reopen the clinic for fear of being charged with yet another crime. The police kept important medical records and lists of activists when they returned the clinic equipment and records to the doctors.

On 7 July 1985 a new JSK clinic was set up outside the UCIL premises. This was run as a general clinic, as the state government refused to continue supplying it with sodium thiosulphate. Dr Ishwar Dasss told the JSK's Dr Vora that he was refusing to supply the JSK with sodium thiosulphate as the clinic was 'not up to standard'. The basis for this objection seems to have been that the clinic walls were not painted and the clinic itself was not spacious enough. Sodium thiosulphate was supplied to the JSK from 9 September 1985, after the Supreme Court had instructed the Madhya Pradesh state to do so. The clinic was run in rooms provided by the Nagrik Rahat aur Punarwas Committee.

Disagreements arose over continuation of the clinic in the Nagrik Rahat aur Punarwas Committee premises, with the committee throwing doubt on the efficacy of sodium thiosulphate and eventually asking the clinic to shift from its premises. While the committee agreed to make its premises available until other arrangements could be made, it did not honour this agreement. According to a statement by the two doctors involved (APPEN, pp. 106–7), they then decided to shift the clinic equipment from the committee's offices. The committee then reported the shifting of the clinic's materials to the police as robbery. Thus the joint initiative among the activist groups on the issue of health care for the gas victims came to an unfortunate end.

No Benefit of the Doubt

A final point needs to be stressed in relation to the medical aspects of the Bhopal killing, a point which is of no small importance to future disasters and present anti-toxic campaigns. In times of crisis, the roles played by various groups become more pronounced and obvious. Bhopal showed the role of government scientists and other experts to be primarily one of reassurance rather than scientific examination of a problem. Thus those medical authorities that were cautious in predicting ill-effects, following the old habit of many toxicologists, especially those in industry and government, of giving the chemical the benefit of the doubt, were incorrect. Those who expected the worst after the disaster, and who were thereby dismissed as alarmists and extremists, have been shown to be factually correct. On the basis of the Bhopal experience, little trust can be placed in the assurances and reassurances of even supposedly neutral medical institutions and doctors.

Without labouring the point, it is important to document some of the areas where the reassuring experts were wrong. On the general health effects, the *New York Times* reported a year after the killing:

According to new medical information the two-hour exposure to gas escaping from the Union Carbide factory in Bhopal caused long-term damage to the lungs, eyes, nervous system, kidney, liver, blood and female reproductive organs. This information contradicts statements in the days after the accident by some prominent Indian and American medical authorities who asserted that there would be few long-term health problems. (1 December 1985)

Assurance about MIC's mode of action also proved to be baseless. The experts said MIC quickly breaks down into non-toxic substances on contact with moisture or water.

One interesting chemical finding [of the ICMR April 1986 summary report] was that MIC was shown to take more than an hour to completely hydrolyse in water. The implication of that finding is that after the leak MIC probably persisted in an unaltered highly toxic form much longer than had been thought possible. Normally it degrades on contact with moisture to relatively harmless methylamine. (Chemical and Engineering News, *14 April 1986, p. 6)*

The experts also said MIC could not enter the body through the skin nor could it enter the bloodstream directly. ICMR studies found MIC was potent enough to enter the body through the skin and can enter the bloodstream directly (*Chemical and Engineering News*, 14 April 1986, p. 6).

In a similar vein, the reassurers claimed there was no possibility of MIC affecting the immune system. Meryl H. Carol told a meeting of the American Chemical Society in September 1985 that MIC injures the immune system (*Chemical and Engineering News*, 16 September 1985, p. 5). The Indian Toxicology Research Centre reported that 55 percent of the 1,109 victims it studied showed a reduction in the immune response (*Chemical Week*, 4 December 1985, p. 12).

While the reassurance experts claimed there was no evidence that MIC would have adverse reproductive or genetic effects, the ICMR found that 366 pregnant women studied suffered 'spontaneous' abortions – 26.7 percent in all, four times the normal rate. To this it must be added that between 35 and 50 percent of affected newborn children were underweight. Also of relevance here is the result of Doctors Bang and Sadgopal's studies, which found an epidemic of gynaecological diseases. As for the fifty unfortunate mice given sublethal doses of MIC in animal tests, according to Dr Y. C. Alarie of the University of Pittsburgh, all suffered spontaneous abortions. Michael D. Shelby of the US National Institute of Environmental Health Science told *Chemical and Engineering News* (3 March 1986, p. 4) that chromosome aberration tests show 'MIC, a breakdown product or a metabolite is capable of producing chromosome damage when applied to cells in culture'. This experimental evidence was confirmed when the Indian Toxicology Research Centre found an abnormally high incidence of chromosome breaks in 31 percent of the gas victims they studied (*Chemical Week*, 4 December 1985, p. 12). Finally, in its suit filed against UCC in Bhopal in September 1986, the Indian government said it is 'reasonably certain' that Bhopal residents were genetically damaged by the gas leak (*Wall Street Journal Europe*, 8 September 1986, p. 7).

The medical cover-up following the killing was based on giving MIC the benefit of the doubt. The studies mentioned above have shown how baseless this was. A conservative desire to limit the health crisis found a strong weapon in the industrial and medical bureaucracy's long history of giving toxic chemicals the benefit of the doubt. Such epidemiological and toxicological research as was undertaken after the killing was generally of questionable or academic value, deliberately biased or kept secret: actual treatment and rehabilitation of the victims was dire. The international medical bureaucracy allied itself with UCC and the Indian state in limiting the crisis. The local medical bureaucracy was split. Concern for the victims was the top priority only of dissident medical and political groups. These groups, which saw the medical cover-up as entangled with the political cover-up, were repressed and marginalized: the state closed down an independent medical treatment centre operated by these groups. All of these tactics had been used before by both state and medical bureaucracies to deal with toxic crises and issues. Bhopal saw perhaps the worst use of these tactics so far.

4 THE PRICE OF CORPORATE KILLING

'The victims are the only innocent people around this case.' – Jacky Hoffinger, court-appointed liaison between the plaintiffs' lawyers in the US Bhopal cases (Wall Street Journal Europe, 9 May 1986)

'If we fail to organize here in India, the safety standards that do exist will probably decline because the ability of a corporation like Union Carbide to weather an incident of this magnitude will set a precedent – a benchmark – for quantifying the risk that companies undertaking these kinds of enterprises are exposed to.

'If they're able to cut a deal – say, $1 billion – and if they're able to raise the money from banks and insurance, and if it's a settlement that they can live with – where they can still make profits – it's likely that the whole community of businesses that deal with these kinds of toxic materials will be encouraged to spend less money on safety because they will be able to quantify their exposure. They will see that they really don't have to guard against the worst because this was the worst, and if Union Carbide is able to survive intact, then companies don't have much to worry about.

'But, if justice is done in this case it will amount to the demise of Union Carbide – the transfer of ownership to the victims. If that were to happen it would send a real message to all the companies that are working with these kinds of toxic materials. If something goes wrong, they stand to lose their corporation and likely the

> *personal fortunes of their managers and directors. That will send a real message and encourage an historic change in the way business is done.'* – Robert Hager (**Multinational Monitor**, *31 July 1985*)

'Only the Victims Lack a Strategy'

Along with the media and chemical warfare researchers, another flock of vultures soon arrived in Bhopal – US lawyers. Smelling the possibility of the largest legal settlement in US court history, US liability lawyers such as John Coale and Melvin Belli descended on Bhopal to sign up clients and then quickly returned to the USA to file billion-dollar claims against UCC. Residents of some areas said US lawyers had offered them 100 rupees to sign up with them. Some of these lawyers got their clients to sign agreements giving lawyers 40 percent of awards, or 50 percent if the case is appealed, with others topping their 40 percent contingency fee with the requirement to pay costs as well. Many of these agreements were in English only, though many of those who signed them are barely literate in Hindi. In some cases US lawyers were reported to have held on to death certificates which they obtained from their clients, thus preventing them from receiving immediate relief from the Indian state and showing the same callousness towards their clients as UCC had previously. One lawyer, John Coale, who persuaded the Bhopal municipality to avail itself of his services, announced to the Indian Press that he had just begun negotiations with UCC looking for $1 billion. The Indian state moved in a hurry also: the Madhya Pradesh government formed a three-person committee to begin negotiations with UCC on 5 December.

There were immediate calls to avoid litigation, with environmental mediation specialists and others calling for the question of compensation to be separated from that of punishment. Thus Robert E. Stein of Environmental Mediation International called for immediate attention to be placed not on who was responsible for the killing, but on the formation of a just compensation scheme. 'What is important is not to apportion blame but to make monetary relief and restitution available.'

While the humanitarian leanings of some of those who made these calls cannot be doubted, it is worth noting that their position was shared by UCC, which also called for an immediate settlement that would be both 'compassionate and reasonable'. Such a settlement would of course be in UCC's interest, for several reasons. It would spare UCC the harassing experience of fighting a long, drawn-out court case, with the attendant publicity and exposure of UCC's negligence. It would also remove the problem of huge legal costs. Such a settlement would also ensure that UCC's full culpability in the killing would be covered up. UCC was also motivated to settle immediately to maintain the image it was projecting as a caring company. And it could instead concentrate on its most recent restructuring plans.

For the Indian government, as well, an out-of-court settlement would have been satisfactory, as it had no desire to see its complicity exposed and documented.[1] Furthermore, there was UCC's threat that the Indian government itself was partially liable for the killing, given that 23 percent of UCIL was held by Indian government trusts. Bud G. Holman, UCC's chief counsel,[2] said that, if the claims were litigated, the Indian government would probably become a defendant as it bore partial responsibility for the accident: 'That raises thorny questions that can be avoided by early resolution.' Thus UCC nudged the Indian government to settle out of court. Another major factor affecting the Indian government position was its desire not to alienate multinational capital and thus compromise Rajiv's pro-high-tech policy. It was essential for the Indian state to be businesslike in its dealings with UCC and not overzealous in attacking the corporate monster, as the Indian state was about to undertake a major attempt to lure multinational capital to its shores. In a situation where 'India probably has made more changes favourable to business this year [1985] than in the past twenty-five years' (Walter Wapel, General Electric's regional manager for Asia, quoted in *Fortune*), hunting UCC too strenuously would have gone against the tide.

Thus both sides agreed on the need to keep the liability question away from open examination and contestation in court. All that remained at issue between them was the price of maintaining silence. One little side-issue needed to be cleared up first: the US lawyers had to be dispatched. This was done in March 1985 when the Indian parliament passed the Bhopal Gas Leak Disaster (Processing of Claims) Act, 1985, a law which named the Indian government as the only recognized legal representative of the gas victims. While the US lawyers have been justly criticized for their greed, their intervention was crucial in beginning legal cases against UCC. 'Without pressure from the Americans, and the possibility of a multi-billion dollar judgement in New York, Union Carbide would have little to fear from any court. It was first the American attorneys who seemed to be conscious about claims and who rushed to the [shanty towns] to take cases,' says Bhopal lawyer N. M. Khan, who assisted Chicago plaintiffs' lawyer Marshall Teichner. 'No other organization or government moved forward for taking claims cases. Several observers note that legal officials didn't get involved until American lawyers embarrassed them into playing a role', Stephen J. Adler reported from Bhopal (*American Lawyer*, April 1985, p. 133). Adler concluded by quoting Marc Galanter, an Indian-law specialist at the University of Wisconsin in the USA: 'Nothing would have happened if it wasn't for the American plaintiffs' lawyers, whatever you think of them. If they hadn't come, the people would have been just left to die, as in so many situations.'

Similarly, while these lawyers' reported cut of 40–60 percent of any eventual damage award is incredibly greedy, at least the victims would eventually receive something. There is little likelihood of the victims receiving much if a deal was made between UCC and the Indian state. One senior lecturer at an Indian law institute told *Chemical Week* (6 March 1985), 'With the Indian government acting in the name of all the victims without any intention of rendering information public, there is a real problem now of how the surviving victims will actually get the money . . . They have no way of controlling the government'; so the government 'can totally hoodwink the people'. Later in the year Dr Misra of Gandhi Medical College told *Business Week* that the government had undertaken no systematic survey of the victims because 'the government wants the whole compensation to themselves to distribute as they like'.

The Indian government also had local reasons for pursuing the Bhopal case in the US courts. The major one was the desire to avoid establishing a legal precedent in India which would badly affect Indian private and state capital in future chemical accidents. In an interview with APPEN Mr Panjwani, a senior advocate practising at the Indian Supreme Court, emphasized this point:

Because the pressure of Indian industry is so great on the Government that if we really set up a tribunal and they set up the principles of awarding compensation on the lines of American law, then the entire Indian law of torts changes. *Therefore in every subsequent case it will change. So it will have far-reaching consequences for which the Indian industry and Indian government were not prepared. (1985, p. 40)*

Similarly, the passing of the Bhopal Gas Leak Disaster (Processing of Claims) Act also prevented independent legal groups in India from intervening in the cases in the Indian courts. The Act also authorized the Indian government to enter into a compromise on behalf of and in the interests of the gas victims.

With the US lawyers safely out of the way, the major question then became the price. The Rand Corporation had recently done a very handy study. According to them, a US life is worth about $500,000. With the appropriate discount, this meant an Indian's life is worth about $8,500. Multiplied by 1,700 deaths, this came to $14.5 million. The 200,000 injured suffered various injuries. US payment to asbestos victims with similarly varied injuries averaged out at $64,000. Adjusted for India this came to $1,100 per person, leading to a total $220 million (*Wall Street Journal Europe*, 20 May 1985). (Another aspect of these cases needs to be pointed out: in asbestos cases from 1980 to 1982, only 37 percent of the awards actually got to the victims, with legal fees and other expenses swallowing 63 percent of awards.)

The major debate became whether payment was to be at American or the discounted Indian values. UCC said the Indian values should be adhered to, while India said US standards of payment should rule. Thus while UCC and the Indian state haggled over the price, the victims and the walking wounded languished. In June 1985, UCIL's Vijay Gokhale said the level of compensation would be 'somewhere between Indian and American standards' (*Financial Times*, 10 June 1985).

As well as haggling over the price, UCC quibbled with the body count. In April 1985, Bud Holman, UCC's counsel, said that the Indian government's figure for deaths was substantially inflated, citing a report to the Indian parliament placing the official death toll at 1,408. Holman also considered the figure of 200,000 injured to be 'definitely on the high side'. Holman told the *Washington Post*: 'Someone in Bhopal that might have smelled gas and felt bad was not significantly affected. There were people who fled their homes; there were people who were *upset* that night. There may have been people who went to first aid because they had felt nausea or felt ill and then left' (Everest, 1985, p. 86, emphasis added). By the end of 1985, some 490,000 people in Bhopal had signed up with the Indian goverment as having been injured in some way by the gas leak.

As UCC toughed out the bargaining stage, its public statements became less full of the milk of human kindness. Their first story was: 'Every effort will be made to mitigate against the deep sorrow of the people of Bhopal' (December 1984). Then: 'The key is to find a way of recompensing . . . that would be compassionate and reasonable. I would hope this could be done in a relatively short period of time' (January 1985). Then: 'We've said right from the start that the proper answer for the people, if you have any compassion for them at all, is not to go through a litigation of liability. We will fight right to the end' (any effort to prove UCC negligent in Bhopal) (February 1985). Then: 'Stockholders should not take our strong interest in achieving a settlement as an admission of legal liability. The corporation did nothing that either caused or contributed to the accident, and if it comes to litigation we will vigorously defend that position' (April 1985). By November 1985 Warren Anderson (UCC chief executive officer) was saying, 'Maybe they, early on, thought we'd give the store away. (Now) we're in a litigation mode. I'm not going to roll over and play dead' (*Business Week*, 25 November 1985, p. 66).

While UCC was prolific in its verbal assurances of its desire to aid the killing's victims, they were in no hurry to put their money where their mouth was. When the case first came to court in the USA, Reagan-appointed Judge Keenan suggested – as a gesture of fundamental human decency, without prejudice to the question of liability – that UCC make an immediate payment of $5–10 million. UCC's response to this request was hardly compassionate.

Holman [UCC's chief counsel] responded by plugging the disaster relief the company had offered already. But he went on to make a curious argument – that to provide further interim relief, Carbide needed specific information about the precise extent of the damage. 'We need to know precisely how many people were killed in this tragedy, how many people were seriously affected, how many people were hospitalized for an extended period, if any, how many were hospitalized overnight, how many were affected by just going to first aid stations . . . The government of India has this information and we need it.' (American Lawyer, November 1985, p. 57)

UCC thus turned a request for a basic humanitarian gesture into a fishing expedition for information.

UCC returned to court two days later offering the minimum figure Judge Keenan had requested, $5 million, but again demanding detailed information. The Indian government refused this offer on the grounds that UCC required detailed reports on each individual receiving portions of the money. The Indian government claimed administrative costs would eat up half the $5 million offered and that report writing would tie up half the working hours of doctors attending the victims. An Indian government spokesperson said, 'Union Carbide just wanted to spend $5 million to buy data [on the number and extent of injuries] from the government' (*American Lawyer*, November 1985, p. 57).

Agreement on the interim relief was finally reached just before the first anniversary of the killing. The money from UCC was to go through the US Red Cross – of which Warren Anderson is a board member – to the Indian Red Cross.[3] The money was to be subject to International Red Cross rules for disaster relief accounting and reporting. However, in a motion by the individual US plaintiff lawyers, filed before Judge Keenan on 21 May 1986, it was alleged: 'To date, only $2 million of this amount has been transferred to the Indian Red Cross and there is no record as to how much has been utilized and for what purpose' (BARC Briefing Paper No. 2, 1986b, p. 3).

When India finally sued in the USA it went the whole hog. India's Bhopal suit included an unprecedented legal claim – multinational-enterprise liability – that 'could strip many multinationals of the insulation that they now have from liabilities of their foreign subsidiaries', according to *Business Week*. The magazine quoted Mark B. Feldman, specialist in international law at Donovan Leisure Newton & Irvine, to the effect that if the claim was accepted 'it would be a major obstacle to international trade and investment as we know it'. If UCC were sued successfully in the USA with the result that US-type damages were awarded, it would be a vitally important precedent that would wipe out the cost advantages of hazard export and become a major handicap for US toxic capital internationally. Nevertheless, this claim may be seen as one more nudge to UCC to settle out of court at a higher figure: US lawyers said that, while India had sued in the USA, it still hoped to arrange a settlement that would avoid the intense scrutiny of evidence and disclosure of documents that would accompany a US suit.

UCC's first offer – reportedly $100 million – was refused. UCC said it 'would be enough to pay heirs of each decedent the equivalent of more than one hundred years annual income and each of the reported serious injuries approximately twenty years annual income in Bhopal'. With Belli and Coale talking of $5 billion damages, this offer was sure to be rejected. Despite its protestations of concern and its desire to resolve the settlement by July, UCC's opening offer was extremely low, less than half its estimated insurance cover. In June the *Wall Street Journal Europe* quoted Asoke Sen, Indian's Minister of Law and Justice, as saying that UCC's offer on Bhopal was only about $100 million. He called media reports of it as $200 million 'highly misleading', as the higher figure represented the value of the offer under a long-term payment plan. Sen rightly rejected the offer as a 'fleabag'. UCC's conditions with the offer included that it would become valid only if all cases against the company were dismissed. UCC was reported in June to have formally offered $230 million, while Indian officials privately said they wanted $1.5 billion. The general consensus reported was that UCC's offer could not rise above $300 million, while for political reasons India could not accept much less than $1 billion.

Robert Hager suggested in the *Multinational Monitor* (31 July 1985) that there was a structural limit of about $800 million, above which UCC could not go. This figure is made up of UCC's insurance and what could be raised by way of bank loans. Going higher would require the management to erode the corporation's assets: 'The management would then expose themselves to shareholder derivative suits against them personally for negligent activity by the subsidiary in India, and that means the personal assets of the directors and managers would be in jeopardy.' This would be of particular concern, as there were reports that UCC had been unable to renew their directors' and officers' liability insurance.

At the same time, in April 1985 at the first hearing before Judge Keenan,

UCC made it clear that they had no intention of abandoning one of their strongest legal tactics: 'If the cases are going to be litigated,' said Holman, 'they belong in India.' UCC was trying to have it both ways, as usual. While UCC argued in the USA that claims should be filed in India, it responded to some of the suits filed in India by individual plaintiffs' lawyers by arguing that only the Indian government could legally represent the victims.

In April 1985, the *American Lawyer* reported that some 2,000 suits had been filed against UCC in Indian courts, mainly seeking interim relief. UCIL was reported to be opposing all these cases: 'Carbide . . . has put together a team of a dozen Indian lawyers – a legal army in a city where almost all lawyers practise solo – to contest every point in the complaints.' Many of the claims were opposed on technicalities such as the plaintiffs' illiteracy: 'The plaintiffs are illiterate and do not understand the contents of the affidavits on which they have placed their thumb prints. Therefore . . . the complaints must be thrown out.'

Lawyers for UCIL were also denying that the Indian company was under the control of UCC, thus denying UCC was in any way liable. They also argued that UCIL was not liable. UCIL's chief lawyer was quoted: 'According to the company, there was no negligence. It was false that Indian engineers did not take care properly. It is absurd to suggest that the machinery was not handled properly' (*American Lawyer*, April 1985, p. 130). UCC was thus fighting every inch of the way in India as well.

UCC's defence that litigation belonged in India could have tied up the US litigation for years. In December 1984 some lawyers predicted that the US suits could take up to five years before the question of compensation was even considered. These predictions undermined one major argument against taking cases against UCC in India: that Indian cases take so long to resolve that many of the Bhopal victims would be dead before the tortuous pace of Indian litigation would come to a final decision. Others – for example, the Bombay Lawyers' Collective, a public-interest law group – argued that the case should be tried in India to create a precedent for future mass-disaster cases.

The two major reasons for filing the suits in the USA were the possibility of higher damages being awarded in the USA and the contention that UCC – rather than UCIL – was the major responsible party. By filing in the

USA, damages could be awarded out of the total assets of UCC, which were large enough to cover even the most optimistic demands. Similarly, by filing in the USA, the lawyers for the Indian government and the individual plantiffs' lawyers could use the discovery process, which allows parties involved in lawsuits access to documents held by each other on the subject of the suit; thus they might obtain material which related to UCC's control of UCIL.

In August 1985 UCC filed for dismissal of the suits in the USA on the grounds that it was an inappropriate forum for the litigation. Since the event occurred in India, most of the material evidence was in India, and the witnesses and the victims were in India, it argued that the litigation should take place in India. UCC further argued that the US plaintiff lawyers did not have proper authority to represent the Bhopal victims. In what seemed an uncharacteristic move for a Reagan-appointed judge, Judge Keenan allowed partial discovery of documents before the forum issue was resolved. Material obtained under this process strengthened the Indian government's claim that major decisions relating to the Bhopal plant were taken by the US parent rather than the local Indian company.

Meanwhile, the negotiations for an out-of-court settlement continued. By August 1985, most US analysts were predicting the maximum settlement would be less that $500 million. What seemed certain was that UCC was unlikely to offer an adequate premium for early settlement of the court case until punitive damages seemed probable. The question of punitive damages, as well as the desire not to pay US-style damages for death and injury, was central to UCC's desire to have the cases heard in India.

Punitive damages are those, above and beyond damages relating to injury and death, which US juries may award in cases of gross corporate negligence or recklessness. Victor Schwartz, a Washington product-liability lawyer, explained the function of punitive damages in the *New York Times* (11 December 1984) in the following way: 'Since we don't have a simple, workable set of criminal procedures that would deter executives from dangerous behaviour that hurts others, we are using punitive damages as a deterrent to protect people. It really should be handled through criminal law, not through the tort system.' It was the threat of punitive damages, rather than the compensatory awards themselves, that forced two major US corporate criminals – the asbestos giant Johns Manville and the Dalkon

Shield contraceptive maker A. H. Robins – to seek protection under Chapter 11 of the US bankruptcy code. Should the Bhopal cases proceed to trial by jury in the USA, UCC's continued existence as a profitable corporation would be under a dire threat.

In the out-of-court negotiations, UCC's delaying tactics were certainly successful. By September 1986 the likely price predicted by the US business press was $600–$800 million, a major climbdown from the original estimates of $10 billion (admittedly optimistic, but also possible, given Texaco's recent shock). When the battle for control of UCC between the present UCC managment and the GAF Corporation (see pages 134–5 below) was in full swing in December 1985, settlement of the Bhopal suits was reckoned to be no major threat to UCC's continued viability. This devaluation was aided by the stated readiness of lawyers for both the Indian government and the Bhopal victims to settle in a short time, rather than confront UCC over a long march with consequent exposure of the full causes of, and various complicities in, the killing. As Robert Hager told the *Multinational Monitor*, 'Bhopal lawyers are less confrontational and more settlement-orientated, questioning whether a strategy of seeking control of Union Carbide's assets is in the best interests of their clients.' Certainly their readiness to deal brought down the price.

UCC was in no rush to complete a deal. At an early December 1985 meeting UCC repeated a settlement bid similar to the $240 million which had been rejected earlier: this offer hardly raised itself to near half of what the Indian govenment eventually expected to get as a minimum payment. The *Wall Street Journal Europe* (9 May 1986) said India had demanded $630 million earlier in 1986. By way of comparison, the *Financial Times* (14 May 1986) cited India's minimum demand as $1 billion. This offer was made when it seemed most opportune for UCC management to have a *fait accompli* on the Bhopal issue to present to its shareholders during its battle with the GAF Corporation.

Of course the crisis of Bhopal was, in UCC management's opinion, successfully handled months before this, before being put on the back burner: it no longer figured as a crisis of the same magnitude, especially when the takeover battle with GAF began to heat up. The slowness of the legal system was obviously in UCC's favour. And UCC intended to utilize this slowness and all the various avenues of appeal to the utmost. In

November 1985 Warren Anderson, UCC chief executive officer, said UCC would take the case all the way to the Supreme Court: 'This could drag on for five or more years,' he noted. Bud Holman, UCC's legal counsel, threatened that UCC would fight each individual case, if the need arose. 'We can litigate 100,000 or 200,000 [separate damage trials] if we want to . . . If we're going to defend ourselves – and we are – if there are 200,000 claimants, the 200,000 claimants are going to have to appear in court' (*American Lawyer*, November 1985, p. 62).

Nevertheless, in early February 1986, UCC made a last-ditch settlement offer before Judge Keenan's ruling on the forum issue, i.e. on which country's courts were to try the cases. UCC offered $350 million, a 46 percent increase on its previous offer. A meeting between the Indian government's lawyers and the individual US lawyers agreed to press UCC for more.

Four and a half weeks later, UCC's offer was secretly accepted by Stanley Chesley and F. Lee Bailey, two of the most prominent lawyers among the individual US lawyers. The intended plan was for Chesley and Bailey to abandon the plaintiffs' management committee, the committee composed of representatives of the US lawyers and the Indian government lawyers, and file a class action suit. This would have given Chesley and Bailey full control of the case against UCC, as well as a large measure of control over and claim on the legal fees incidental to the settlement. Bailey and Chesley's controlling position was jolted slightly when UCC forced them to include in their new panel two Philadelphia lawyers who were bringing cases against UCC in the state courts: UCC insisted on this as it wished to end all state, as well as federal, court cases against the company in the USA. US chemical industry analysts were surprised by the low settlement figure when it was announced. Most had expected it to be in 'the $500 million to $700 million range' (*New York Times*, 24 March 1986, p. D1).

Reaction to the announcement of the deal was swift. The Indian government's US lawyer denounced it as a sell-out. Stanley Chesley publicly praised it. Two US lawyers, John Coale and Arthur Lowy, placed an advertisement in a Bhopal newspaper informing the readers the plan would result in the payment of around $25,000–$66,000 per death claim. The settlement was denounced by the Indian government. Keenan dismissed the settlement on the basis that any valid settlement would require the involvement of the Indian government. In May 1986, Judge Keenan finally announced his judgement: the cases were to be transferred back to India.

Getting the Capital Out

One major criticism of Judge Keenan's decision was made by US public-interest lawyer Robert Hager: 'If Judge Keenan was going to decide this, he didn't have to wait over a year to do it. He gave Union Carbide time to divest itself of its assets.' Thus UCC was able to get its capital out while the court case was in progress in the USA. In December 1985 UCC set out a major restructuring programme which led to a massive reduction in the assets UCC possessed from which to pay the Bhopal claims. While this restructuring was finally forced on UCC by its battle for management control with the GAF Corporation, the net result was the same. A large part of UCC's capital base was sold off, major assets such as the consumer products division were divested and the proceeds distributed as a special dividend to the UCC shareholders; UCC also bought back stock from its shareholders.

'I was surprised that the courts let Union Carbide go through with this reorganization', Paul Christopherson, chemical industry analyst with investment brokers Bear Stearns & Co., told *Environment Action* (September/October 1986, p. 10). Christopherson cited the example of the A. H. Robins company, the maker of the Dalkon Shield IUD contraceptive, which also faced potentially ruinous claims against it and which was under strict orders not to divest itself of any of its assets. Christopherson's analogy is inappropriate, however, as the critical difference between UCC and A. H. Robins is that A. H. Robins filed for the protection of the US courts under Chapter 11 of the US bankruptcy code. This allows A. H. Robins the protection of the court while it makes reorganization plans and plans to meet the avalanche of claims against it. While the Chapter 11 filing allows Robins protection from court cases until its reorganization plan is completed, it also means control of the company passes to the courts.

UCC had not placed itself under the protection and control of the US courts. While Judge Keenan may be criticized for allowing UCC to restructure by selling off a major part of its asset base, the blame for allowing the restructuring to proceed unchallenged lies with the US lawyers and the Indian government. Both the lawyers and the Indian government are much more open to criticism than is Judge Keenan, as neither filed to block UCC's restructuring, though threats to do so were made once or twice. In the case of the individual US lawyers, this is no major surprise. Their conciliatory and co-operative approach to UCC, on the basis that the only way they would get their fees was through a negotiated agreement with UCC, meant that they could not threaten UCC in this manner. Similarly, the Indian government did not challenge UCC's restructuring for the same reasons it had not attacked UCC too strenuously: it feared the effect on its recent policy of inviting multinational capital into India.

While UCC planned a major restructuring after Bhopal, it was the battle for management control with the GAF Corporation in December 1985/January 1986 that forced it into the major restructuring it finally undertook. This take-over bid and the jockeying for control of UCC added a bitter twist to an already bitter story. It provided us with a sickening example of the vile perversity of the owners and controllers of toxic capital in the USA. While two vultures fought over the carcass of UCC, one was playing tough in its negotiations with the Indian state over Bhopal and the other had no compunction about using the Bhopal issue as part of its take-over strategy. Thus Samuel Heyman, the chief executive officer of GAF who spearheaded the take-over bid for UCC, held up to UCC shareholders the illusory scenario that he would move quickly to resolve the Bhopal suits as part of his take-over strategy. Though Heyman was reported to have put a high priority on settling the Bhopal suits, he had previously shown himself to be an extremely tough litigator. In December 1983, GAF had 12,000 asbestos-related suits pending against it. T. W. Henderson, who had fought GAF in asbestos-related court cases, noted that, 'GAF fights to the bitter end.' *Business Week* pointed out, 'It would be to Heyman's advantage to have Carbide's shareholders think he would move towards settlement more quickly than current management.'

During the battle with GAF, UCC handed out golden parachutes to its golden boys far more lavishly than it gave relief to the Bhopal victims. UCC directors had allowed, in the event of a successful hostile take-over $28 million for some forty-two top executives, $8.7 million of which was for its top five officials. Neither result of the battle boded well for the victims of Bhopal: Heyman intended funding his bid with high-risk junk bonds and intended selling off choice assets to pay for the take-over. UCC finally defeated Heyman by pursuing his strategy themselves, announcing their intention to sell their consumer-products division, which accounted for $1.9 billion of UCC's $9.5 billion 1984 sales – a particularly profitable part of total sales. The cost of compensating the victims of Bhopal pales compared with the debts incurred in fighting the take-over, the legal and banking fees and golden parachutes.

Beating off GAF doubled UCC's debt to $4.5 billion and slashed its equity value to a quarter of what it had been. The new, highly leveraged, UCC was reliant on 'highly cyclical industrial businesses' with a 'low-growth mix of industrial gases, chemicals, and plastics'. GAF came out of the battle with a gain of $200 million. After the battle, GAF's merchant bankers and lawyers were owed $60 million. UCC's bankers took at least $14 million, while Anderson complained, 'Wall Street is becoming a casino rather than an investment organization.' UCC was forced into taking these moves by the changing composition of its shareholders: by December 1985, an estimated 30 percent of UCC stock was in the hands of take-over speculators and arbitrageurs (institutions, according to *Forbes* magazine, recently described as the 'house magazine of the rich', that would sell their mother if the price was right). Ownership by banks and other financial institutions sank to 35 percent from 60 percent in late 1984.

The result of UCC's restructuring was a company whose base had been drastically curbed. R. S. Wishart, UCC's vice-president for public affairs, said in August 1986 that UCC's divestitures for 1984 to 1986 would reduce sales over 20 percent to $6.8 billion, while cutting assets by 27 percent and staff by roughly 50 percent. *Business Week* (2 June 1986, p. 52) estimated UCC's assets had been cut from $4.9 billion at the end of 1984 to $697 million at the end of 1985. At the end of March 1986, UCC's debt was $5.5 billion, eight times its equity. Some $2.5 billion of this was in the form of debt paying high interest, varying from 13.5 to 15 percent over periods of seven to twenty years. In the first nine months of 1986, UCC's interest bill reached $410 million, double the interest for the same period in the previous year (*Business Week*, 17 November 1986, p. 49). This nine-months' interest bill was higher than the highest offer UCC had made to settle the Bhopal suits. The obscenity of these calculations is beyond comment.

Back to India

In May 1986 Judge Keenan finally settled the forum issue by sending the Bhopal cases back to India. He justified his decision on unusual grounds: any other decision, he said, 'would be yet another example of imperialism' and would 'revive a history of subservience and subjugation from which India has emerged'. He attached three significant conditions to his judgement: UCC must agree to submit to the Indian courts' jurisdiction; UCC must pay any damages awarded by the Indian courts; UCC must agree to participate in pre-trial discovery proceedings under US procedures, which are much broader than similar Indian legal procedures.

Both sides claimed the judgement represented a victory for them. India's US lawyer, Michael Ciresi, told the *New York Times* (13 May 1986, p. D1), 'These conditions are the primary reasons why we came to the United States courts. We wanted to be sure Union Carbide's assets were subject to any judgement we got and that the company submit to pre-trial discovery proceedings.' In fact the judgement represented a partial victory for both sides. UCC got its wish to have the cases heard in India, though under more onerous conditions than it would have wished for. Having originally sought to have the cases heard in India, UCC were in a bad position to refuse the conditions Judge Keenan had imposed. Nevertheless, UCC requested sixty days to decide whether to accept Judge Keenan's ruling. UCC was supported in this by the individual US lawyers, who still hoped an out-of-court settlement would be agreed in that period.

On 10 July 1986, UCC appealed against that part of the ruling that said only UCC would be subject to US pre-trial discovery procedures. Observers noted yet again that UCC as usual wanted it both ways. Having fought the case in the USA on the grounds that the Indian legal system was adequate to deal with the case, UCC was now demanding discovery rules that were over and above those provided by the until then allegedly adequate Indian legal system. Warren Anderson justified UCC's application for the Indian government to be bound by the same discovery rules by contending that, 'This is important because the Indian government has not been forthcoming with information in the past.'

Keenan also gave UCC a major weapon with which to defend itself and, if it wished, to drag out the legal process to the extent possible. Judge Keenan stipulated that the Indian case must comply 'with the minimum requirements of due process'. This gives UCC major grounds for appeal before US courts against any enforcement of a judgement made by Indian courts. Thus, if awards made in India are at US levels of compensation, UCC can argue this violates due process. Indeed UCC can and probably will argue on various violations of due process *ad nauseam*.

One immediate example was shown in its attitude to interim payments. Indian courts have previously ordered interim payments to be made to various plaintiffs while cases are still pending. But a UCC spokesperson told *Chemical and Engineering News* (30 June 1986, p. 10) that this occurred only when a defendant had admitted liability and was contesting the size of the reward. (It will be remembered that UCC had not admitted liability, only moral responsibility.) 'That's not true in this case. If interim payments were ordered, we believe they would constitute a violation of due process.'

Generally the decision was welcomed by public-interest groups. Dembo and Morehouse (1986, p. 3) saw advantages in the ease of access to the courts in India, the reduction of lawyers' fees and possible court orders to prevent further divestment of UCC assets and to order UCC to provide interim relief. They also cited a recent case, to show that the criticism of dilatory Indian courts might not apply to the Bhopal case: 'The handling of the December 1985 Sriram factory gas leak in Delhi comes immediately to mind. The Indian Supreme Court asserted direct jurisdiction through a citizen petition filed three days after the accident, and by the following April, had largely dealt with the compensation and public-safety issues involved.'

In the run-up to the trial, both sides jockeyed for advantage. While the trial's beginning was inevitable, it was obvious that both UCC and the Indian state still wished for an out-of-court settlement. Thus many of both sides' actions over this period were intended to nudge the other side to strike a bargain. UCC's restatement of the 'sabotage' theory should be seen in this way. Even if sabotage was proved, UCC would still be liable for damages due to the doctrine of strict liability, though evidence of sabotage would give UCC an opportunity to complicate and prolong the Indian legal process. *Chemical and Engineering News* reported:

> *Evidence of a deliberate act could complicate trial proceedings, however, experts say. 'It won't exculpate Union Carbide, but it offers great possibilities for dragging things out,' remarks Marc Galanter, an Indian-law expert at the University of Wisconsin law school ... Carbide, for instance, could request a stay of the civil case until a criminal trial of a suspect was completed and, perhaps, appealed. (18 August 1986, p. 4)*

Futhermore, if the Indian courts decided Carbide's theory was not adequately backed by factual evidence, UCC would be able to claim it had been denied due process, thus providing itself with another basis for appeal before US courts.

The possibilities for dragging out the court case are immense. Indian sources told *Chemical and Engineering News* (15 September 1986, p. 14) that a final judgement would take at least five years and possibly fifteen. As Galanter pointed out to *Chemical and Engineering News* (18 August 1986, p. 4), 'Indian civil procedures give great scope for interlocutory appeals – appeals of specific points made in the midst of proceedings that must be decided before the case can go on. If India, mindful of the due-process condition, does not restrict such appeals or route post-trial appeals directly to its Supreme Court, delays could be extensive.'

While on the one hand UCC was showing how it could delay the Indian trial, on the other hand it was indicating its preparedness to settle out of court. At a conference on industrial-crisis management in New York at the beginning of September 1986, UCC's Warren Anderson again stressed that UCC was not interested in a protracted trial:

> *My feeling about litigation was stated from the word go. If you want to litigate victim liability, I'll take as long as it takes to do it. But I want to avoid that issue. I accept moral responsibility. Our name was associated with that operation. People were hurt. Let's solve that problem. Let's not get caught up in the morass of litigation and responsibility because an individual might want to know fifteen years from now what exactly did*

happen and who was to blame. (Chemical and Engineering News, *15 September 1986, p. 15)*

Thus Anderson reiterated UCC's original position and showed his desire to deal. Sources quoted by *Chemical and Engineering News* claimed there was no major disagreement over the amount of settlement between the two sides: India would accept, they said, $500 million spread over seven to ten years. All that was necessary was an acceptable mediation process.

The Bhopal case began in the Bhopal court on 6 October 1986. It was adjourned when no UCC representative appeared in court. The first petition filed in the case was from the Indian government asking the court to order UCC and UCIL not to interfere or tamper with evidence in the case. Meanwhile, out-of-court settlement attempts continued. A team of Indian officials left for the USA on 8 October 1986. With no result from these negotiations, it was now India's turn to put pressure on UCC. On 4 November 1986, UCC announced plans to buy back the $2.5 billion in high-interest debt that it incurred in fighting off the GAF take-over attempt. It proposed to do this by borrowing more than $3.1 billion and offering a premium of 15–33 percent over the debt's original purchase price to ensure success (*Business Week*, 17 November 1986, p. 49). On 17 November, the Bhopal court issued a temporary injunction blocking UCC's recapitalization plan after the Indian government had argued that UCC was placing its debtor's interest before the interest of the gas victims. The Indian government also increased the stakes by asking for $3 billion in its claim to the Bhopal court.

UCC's counter-claim attempted to transfer liability from the company to the Indian state and federal governments. UCC argued that the Indian government had forced UCC to produce MIC at Bhopal, that the government was aware of the inherent dangers and, despite this awareness, had legalized temporary dwellings near the plant. The company also used its reliable objection, arguing that the temporary injunction did not comply with due process: 'To attempt to require UCC now, as the price of defending itself, to be subject to an injunction that prevents it from improving profits, cash flow and net worth, makes a mockery of the due process of law' (*Chemical Marketing Reporter*, 24 November 1986, p. 3). UCC further argued that Indian civil law forbade a court to restrain property outside India. UCC offered, however, to maintain $3 billion in unencumbered assets to cover any judgements, if the court agreed to lift the injunction. On 31 November the Indian court lifted the injunction on asset sales, having ordered UCC to maintain $3 billion to cover the claims. UCC reacted to the Indian goverment's $3 billion claim by filing an affidavit saying the Indian government had been previously prepared to accept $630 million in an out-of-court settlement.

More than three years after the killing, the victims of Bhopal have yet to receive any compensation for the injury done to them by UCC that fateful night. Some critics will argue that the litigation that followed Bhopal was not in the victims' best interests. The major problem with the legal moves, they say, is that no immediate cash payments were made to provide immediate relief to the victims. The length of time such litigation takes ensures the victims' suffering will continue for years without relief or compensation. This is undeniable. Yet a failure to litigate against UCC would have meant the victims would have been dependent on what charity UCC saw fit to provide. UCC's preliminary aid offering, in the week after the tragedy, was $1.5 million. When the first cases were filed against UCC in the USA, UCC immediately argued that what was required was not to fix the blame but to provide money to ameliorate the victims' suffering. Thus UCC argued against litigation on the basis of compassion for the victims.

Yet the compassion of a multinational corporation has strict monetary limits. UCC's offer of $100 million in legal settlement was minuscule in comparison to the expenses involved in relieving and rehabilitating the gas victims. By way of comparison, the US Citizens' Commission of Bhopal has estimated some $4.1 billion in damages are required for restitution of economic losses alone, with possible damages of $15–$20 billion involved if non-economic damages for pain and suffering and punitive damages are included (Morehouse and Subramaniam, 1985, pp. 62–7).

UCC's compassionate concern for the victims was nothing but a cloak for its own economic interest. Its 'compassionate' stance was belied by its failure to offer interim payments and by its later opposition to the possible ordering of such payments by the Indian courts. The choice was not that between immediate out-of-court compensation or waiting years while the litigation wound its way through the courts. The former choice was no choice, as the immediate compensation offers were so miserly as not even to approach basic needs.

It is undeniable that the legal system does not provide an adequate means for victims of toxic catastrophes to obtain redress. Yet it is also undeniable that it is one of the few means available to the victims of toxic catastrophes. By what other means could the victims of Bhopal obtain relief and rehabilitation from the corporation responsible for their sufferings?

The American lawyers who flocked to Bhopal have been strongly condemned; yet, as Adler pointed out, they are the ones that began the only process that was likely to obtain some adequate redress for the victims.[4] Without the intervention of the American lawyers, the people of Bhopal would be in the same position as the people who suffered from the previous toxic disasters of 1984 at Ixhuatapec, Mexico City and Cubatão, Brazil: forgotten. Not only forgotten, but also powerless. While the victims of Bhopal were able to organize against the Indian state, the litigation was the only way in which they could struggle against UCC. Undoubtedly control of the litigation was not in their hands: as the *American Lawyer* noted, the only ones without a legal strategy were the victims. The intervention of the Indian state, with its arrogation to itself of the right to sue UCC on behalf of all the victims, makes it doubtful that much of the final settlement will trickle down into the victims' hands.

Yet the litigation was essential, if not for the individual benefit of the Bhopal victims, then certainly for the benefit of other potential victims. Gersuny (1981, p. 8) points out that 'litigation is a form of class conflict'. The litigation against UCC was essential to prove to the chemical industry in the only terms it recognizes – dollars and cents – that toxic killings are not cheap and even poor Indian labourers are not expendable. That the litigation has degenerated into little but haggling between two crooked partners over the price of mass murder is no surprise. Nevertheless, it is essential that the price extracted should be the highest possible.

PART TWO
MORE BHOPALS

5 Fall-out from Bhopal

'It is likely that Bhopal will become the chemical industry's Three Mile Island – an international symbol deeply imprinted on public consciousness. Just as Three Mile Island spurred a thorough assessment of the safety of nuclear power, Bhopal will bring justifiable demands that hazardous facilities in the chemical industry be designed, sited and operated so that nothing even close to Bhopal can ever happen again.' – James Speth, president, World Resources Institute, and chairperson of the US Council on Environmental Quality under President Carter, 1985

Bhopal represented a crisis of potentially major proportions for toxic capital. In the first place, the killing was a major blow to the image of a caring, safe, socially responsible industry – an image that the industry had spent much money and energy in promoting. Furthermore, it raised a new major issue relating to the immediate safety of communities surrounding chemical factories and storage areas, an issue which had not been raised since the Flixborough explosion in England in 1974. As Ronald Lang of the US chemical-industry association, the Synthetic Organic Chemical Manufacturers' Association, expressed it:

Bhopal focuses concern on something that had not been adequately addressed before – the possibility of catastrophe. Chronic health concerns had already been in the public mind, but Bhopal tells us that while you may have been concerned about what exposure to a chemical will do twenty years down the road, now you have to worry about your community being wiped out overnight.

William Stover, of the US Chemical Manufacturers' Association (CMA), has attributed the muted public reaction to Bhopal to four factors: the speed and competence of the UCC response, a negative public reaction to the descent of US liability lawyers on Bhopal, the good safety record of the US chemical industry, and the assumption that such an accident could not occur in the USA, or the developed world in general. Toxic capital was extremely lucky that the killing did not take place in the USA but in the Third World. The crisis could have been deepened if continuous reports of leaks were reported in the mass media in the same way they covered a Swedish leak two days after the Bhopal killing. However, the industry didn't keep up the body count. Bhopal then became just one more of the many images of disaster in the Third World in 1984 that provided staple fare for the sated consumers of the Western mass media.

It was essential for capital to limit the safety crisis to the Third World. Some capitalists wished to limit the problem to the Indian subcontinent itself: they attacked in particular the Foreign Exchange Regulation Act, which diluted multinational capital's control of joint ventures in the subcontinent. Inside the strategy of limiting the crisis to the peripheral countries, it was also essential that the crisis be limited in time. The acute effects of the gas release were so horrible that it was essential for capital that there should not be chronic effects, or at least that these effects should be minimized or go unreported: this was necessary both to play down the danger that the industry represented to the public and to maintain a division between those fighting the acute and those fighting the chronic effects of toxic industry. This particular portion of capital's response was handled by the UCC-arranged chorus of reassurances. Nevertheless, the industry had to mobilize itself to deal with the general crisis.

Peripheral Countries

Perhaps the most obvious example of the double standard in relation to toxic industry worldwide was shown by the legislative response to Bhopal. While the USA began major Congressional inquiries into the chemical industries and various members of Congress began a race to see who would regulate the chemical industry first, in the Third World little change was expected and little arrived. Such actual measures as were initiated by peripheral countries' governments were focused solely on methyl isocyanate and pesticides production.

An immediate response of sorts was inevitable in India. The Gujarat Pollution Control Board ordered that twelve pesticide plants be closed pending a review of their safety precautions. On 5 December 1984, the Arjun Singh state government in Madhya Pradesh announced a survey of all industrial plants in the state. Its aim was to investigate whether the potential for a similar catastrophe existed elsewhere in the state. The states of Maharashtra and West Bengal also ordered their Pollution Control Boards to study the safety standards of all potentially hazardous factories. The national government also announced an offer of tax sweeteners to all firms which installed pollution-control equipment.

However, this flurry of activity resulted in no major changes. The Indian media became far more alert in relation to toxic industry. Toxic leaks at various factories were covered at length, while they would have been ignored before. Environmental and political groups also began to take up toxic issues: 'There is a great deal of organized opposition to the siting of hazardous factories. That's true of Kerala, Gujarat, Karnataka' (Bidwai, 1986). However, this upsurge in opposition to toxics was not strong enough to force the Indian state to act decisively to restrain toxic industry.

This failure may be attributed to two factors. The less important factor was the lack of trained staff. More important was the state's desire not to upset toxic capital too greatly. The Indian state had just embarked on a major attempt to woo multinational capital to its shores. It did not wish this attempt to be compromised by tightening safety regulations.

In the immediate aftermath of the killing, the president of the Indo-

American Chamber of Commerce told the *Wall Street Journal:* 'I don't think India is going to worry too much about multinationals as a result of what happened in Bhopal, except that in the case of dangerous processes, we probably will want to get some safety commitments.' The value of such commitments by multinationals can be seen in the fact that UCIL's licence for the Bhopal plant was renewed by the Indian state in 1982 on the basis that UCC would make available the most modern technology to deal with safety and hazards. Of course, the Indian state was also under pressure from the local chemical industry, including the major public-sector chemical companies, to minimize its regulatory response.

Two examples of the actual behaviour of the Indian government in relation to toxic industries illustrate how little or nothing had changed in the government attitude. The first is an anecdote related by Anand Grover of the Lawyers' Collective, Bombay:

There is one of the biggest fertilizer plants in Bombay . . . it was producing ammonia for its fertilizers and one of the ammonia plants was very bad and the other was fairly good. There was an ammonia leak from the bad plant but the notice, the stop work notice, was issued in respect of the good plant. They just didn't find out where the leak was coming from. (Jackson et al., 1985, p. 14)

The second example relates to the Sriram Food and Fertilizers Industries plant in Delhi. On 4 December 1985, a release of oleum from this plant led to the deaths of seven people and the hospitalization of some 500 people. A committee set up after the Bhopal killing to investigate this plant had concluded that its relocation was necessary. On the basis of this report the Delhi Administration Labour Commissioner had called on the central government to close the plant and relocate it. In a direct echo of Bhopal, the authorities responded by relocating the Labour Commissioner. Even after the deadly leak, the authorities attempted to minimize the threat: 'True to its colours, the Administration added to the confusion by first suppressing the news and then attempting to minimize the impact of the incident. Dr Varadarajan, the director-general of the Council for Scientific and Industrial Research, stated that oleum was not harmful – which is contrary to the evidence contained in standard textbooks on the subject' (*People's Democracy*, December 1985). The plant was finally closed by the state only when a citizens' group petitioned the Indian Supreme Court to do so.

> ## Toxic Leaks in India, 1985
>
> **January:** More than one hundred people injured when sodium hydrosulphate leaked from a warehouse in Jabalpur.
>
> **6 January:** Forty-five workers treated for exposure to a chlorine leak from Madurai Coats factory in Kerala.
>
> **1 April:** Three dead and two injured in a leak of sulphur dioxide from a chemical plant in Bombay.
>
> **29 June:** A chlorine leak from a drum on a truck injured 113 people in a Bombay suburb.
>
> **August:** Thousands of people affected by a gas leak from a chemical factory in Kanpur, Central India.
>
> **19 August:** 125 people injured after a chlorine leak from a factory in Gwalior, Madhya Pradesh.
>
> — Thirty workers affected by a leak of unidentified gas at a factory in Kalyan, Maharashtra.
>
> **30 August:** One death and 149 people injured when chlorine gas leaked from the Calico Mills plant in Chembur, Bombay.
>
> **September:** 200 people injured by a chemical leak from a lorry in Cochin, South India.
>
> **16 October:** Twenty-nine injured in a chlorine leak from a Bombay pesticides plant.
>
> **22 November:** Three people killed and 200 injured by a leak of an unnamed gas at a factory on the Noida Industrial Estate, New Delhi.
>
> **4 December:** One person killed and 350 injuried after oleum leaked from the Sriram Food and Fertilizers plant in New Delhi.
>
> **6 December:** Sulphur dioxide gas leak from the Sriram Food and Fertilizers plant in New Delhi.
>
> **21 December:** Five people killed and forty-five injured in an explosion and leak of acitol at Somaiya Organo Chemical Company's distillery in Ahmadnagar.

The reluctance of the Indian state to deal with the issues Bhopal raised is confirmed by other observers. The Centre for Science and the Environment in its annual report on the Indian environment concluded: 'Five months later, the government still has not taken any steps to prevent the recurrence of Bhopal-type disasters. A proposal to set up an inter-ministerial board on hazardous chemicals had been quietly buried . . . What is clear, however, is that the government has not even thought of developing any emergency response system to industrial disasters.' The local journalist who had predicted the killing observed, 'I do not see any change . . . All I see is a desire for more of the same.' On the anniversary of the killing, the *Wall Street Journal* concluded another Bhopal could happen again in India. *Newsweek* wrote (14 November 1985), 'The world's worst industrial disaster, it seems, wasn't enough; it may take more wholesale deaths to jolt New Delhi and the multinational corporations into cleaning up India's chemical wasteland.'

In other peripheral countries, the response was desultory. When any regulatory action was undertaken, it was specifically focused on methyl isocyanate. In early February 1985, Taiwan ordered the suspension of all imports of sodium cyanate, used by six pesticide manufacturers to produce methyl isocyanate. The suspension was ordered following a government investigation which found safety measures and emergency equipment at plants using sodium cyanate inadequate. In Brazil, two key ports were closed to the transport of methyl isocyanate. This created problems for three US companies operating in Brazil. One of these was none other than UCC.

The varied response to Bhopal by US companies in Brazil is interesting. American Cyanimid halted production of Spike, a herbicide using MIC as an ingredient, which it produced in Resende for Elanco, Eli Lilly's agricultural operation. This was done after the local plant manager requested the factory be closed until more was known about Bhopal. By way of comparison, UCC wished to continue production in its plant in highly polluted Cubatão. UCC engaged in a publicity battle over whether a shipful of MIC should be processed at the plant or returned to the USA. UCC lost and the MIC was shipped back to the USA. Union Carbide de Brasil spent $2 million on pollution control. Following a long-term ban on the transportation of MIC, Union Carbide de Brasil switched to a new production process using an unspecified intermediate chemical.

Interestingly, during a major controversy over the safey of UCC's French subsidiary La Littorale's plant in Béziers in 1978, UCC told local ecologists that it was impossible to manufacture Temik using an alternative process that would not require MIC stocks to be held. The truth of the matter was that stocking MIC was an economic decision. If MIC isn't produced at one central plant and then transported to be used at various other plants, there is a need for increased capital investment to produce MIC at each individual plant.

Aside from these immediate responses relating to MIC itself, little changed. A Chilean Foreign Ministry official told the *Wall Street Journal Europe*, 'If a foreign company wants to invest in Chile, there are very few obstacles we will put in front of it. And that includes environmental risks.' A multinational manager in Chile told the *Journal* that health inspectors could be persuaded not to inspect factories for the price of a lunch and a couple of beers. After the Bhopal disaster, Merck, Sharp & Dohme held a meeting of its factory workers in Ireland to reassure them a Bhopal-type accident could not occur at their plant, but there has been no reported surge of multinational corporations informing the nearby public of the hazards their plants pose. A year after the killing, the *Wall Street Journal* reported that most Third World countries have been relying on the multinationals themselves, rather than state regulation, to prevent another Bhopal. Yet the chemical-industry press reports that multinationals have tightened safety procedures, upgraded plants, increased supervision and maintenance and undertaken worker and community education programmes.

Metropolitan Countries

In the metropolitan countries, Bhopal represented a major set-back for toxic industry. In both the USA and Britain, an alliance between the chemical industry and reactionary political leaders had recently succeeded in pushing through policies of deregulation and self-regulation. These policies were more successful in Britain than in the USA, where the Reagan regime shot itself in the foot with Sewergate and various other scandals.

(The Sewergate scandal involved 'accusations of political favouritism, sweetheart deals, conflict of interest, misappropriation of funds and the destruction of evidence' against the Environmental Protection Agency, which led to the removal of the Agency's head and twenty-one other political appointees.[1])

In the USA the industry quickly recognized that Bhopal represented a common problem to which a common response was necessary. Thus H. J. Corbett, Monsanto's vice-president for environment, said, 'It would be terribly inappropriate to attribute this problem only to Union Carbide. Union Carbide is a fine, socially-responsible company. In that sense, if this sort of thing could happen to Union Carbide, one has to operate as though it could happen to any other company.'

While the industry was to respond as a whole to Bhopal, individual companies broke ranks in the immediate aftermath of the killing, specifically those companies using MIC. Bayer AG, the only European company manufacturing MIC, produces it at two sites, Antwerp and Dormagen, both of which are large centres of population. Bayer annnounced that it used an entirely different process to make MIC, involving non-toxic intermediate products and a less volatile end product. It also stated that it produced only small quantities for immediate use. Unlike Bhopal, Bayer did not store MIC under pressure: it kept it below ground in refrigerated stainless-steel tanks with double walls and sophisticated monitoring equipment. Bayer's actions were in response to the call by the German Greens for a freeze in the production and transport of all carbamate pesticides in which MIC is used.

In the USA, Du Pont, FMC and Morton Thiokol all distanced themselves from UCC by stating that their MIC storage procedures were different from those of UCC at Bhopal. FMC said it stored MIC underground in stainless-steel cylinders inside a concrete-wall vault which is constantly monitored. Tanks are never more than 60 percent full. Air pressure in the vault is reduced and MIC is kept at 32°F to retard vaporization. Du Pont, which uses 3.4 million lb. of MIC annually, stores MIC in two 300,000 gallon tanks, which are kept above ground for easier inspection. The tanks, which are 60 percent full, are refrigerated. Both FMC and Du Pont storage systems are vented to incinerators in case of a gas leak. The Japanese also broke ranks, with Mitsubishi describing MIC as 'too dangerous to store in a tank'.

Despite these efforts by individual companies, the industry needed to formulate a general response. For the USA, Warren Anderson of UCC said the Chemical Manufacturers' Association (CMA) would spearhead the industry's evaluation of the Bhopal fall-out:

> *I think you're going to see a banding together to address the issues raised by Bhopal. I can cope with what's inside my fence. Outside of it it's an industry problem . . . I think what's going to happen is an appraisal by the multinationals of Japan, Europe and the USA. We have a new issue here as to how far to go in bringing high tech into developing countries.*

The general industry response was to say that chemical sites in the metropolitan countries were safer than those in peripheral countries, as they did not rely on unskilled labour, have far more sophisticated emergency warning and response systems and are part of a 'culture of safety'.[2] Industry spokespeople emphasized that the systems in use are the safest that can be designed but noted accidents are always possible, as human error cannot be controlled. The racist implications of the first part of this argument cannot be ignored. One US UCC employee was quoted by *Business Week* (25 November 1985, p. 68), 'We're bitter. A few, incompetent, casual Indians put a black mark on my name.'

In Europe, the response to Bhopal did not reach anywhere near the crisis point it reached in the USA. Both state and capital reassured the public that everything was under control, citing the EEC Seveso directive on major industrial hazards as sufficient regulation to protect them. This reassurance was, like so many others, just that. The Seveso directive was intended to be fully implemented by all EEC states by 8 January 1984. But by February 1986 fewer than half of the EEC states had even formally implemented it. This was par for the course: only one of sixty-odd EEC environmental directives has been implemented on time and correctly; that directive, limiting mercury discharges by the chloralkali industry, only applied to four of the twelve member states of the EEC. On 13 December 1984 the European parliament, an impotent talking-shop, called on the EEC Commission to ensure that EEC firms observe the same safety standards in Third World countries as they do in Europe. Later in 1985 the

European Council of Chemical Manufacturers' Associations produced a statement of principles on technology transfer. This stated that the standards of safety should be the same in Europe and the Third World, but that the technical means to ensure that safety standard may vary from country to country.

In Britain the Health and Safety Executive was reported to be 'crawling all over' chemical plants in a sudden excess of regulatory zeal. Ciba-Geigy, the only company in Britain to use MIC, suspended its use for five weeks and joined the chorus of companies strenuously pointing out the differences between their storage practices and those at Bhopal. The executive-council chairperson of the Transport and General Workers' Union derided assurances that Britain has tighter rules on toxic chemicals than India: 'These safeguards have been systematically whittled away by the Tory government to the point where the factory inspectorate is too stretched to enforce regulations.' The health and safety officer of the General, Municipal, Boilermakers' and Allied Trades Union pointed out: 'It's ten years since Flixborough, yet our major hazard laws won't operate until 1989.' He called on his union's members to co-operate with community groups and councils to minimize the risk posed by toxic chemicals. Despite this rhetoric, no major campaign was launched by either trade unions or environmental groups to force the publication of a full list of factories affected by the Seveso directive.[3] This publication in no way threatened the Chemical Industries Association Ltd's previous victory over the Health and Safety Executive regarding the number of factories that would be publicly named under the provisions of the Seveso directive (see Appendix 2).

The general lack of response to Bhopal in Europe was due to the weakness of anti-toxic opposition there before Bhopal. This opposition had been confined to opposing individual plants and dumping operations, rather than developing a general attack on toxic industry. Furthermore, the major Green and environmental groups were concentrating their efforts on the nuclear issue, both civil and military, and the question of acid rain.

A measure of complacency in relation to toxic chemicals was also involved. This complacency was also shared by the European chemical industry. Speaking in November 1985 in London, D. R. Bishop, from the US multinational Monsanto, warned the European chemical industry that legislators might extend regulation of the industry in Europe 'unless industry is clearly seen to be meeting the public need for more information [on toxic industry's operations] and better emergency planning'. At the same conference L. Jourdain, of the European Council of Chemical Manufacturers' Associations, complacently explained that the industry accepted and even welcomed government regulations, as they 'give the public some sense of protection, curbing its aspirations to the right to know, which is then no longer necessary for bolstering [the public's] claims for compensation' if a chemical disaster occurs. This complacency was to be shattered a year later by the release of toxic gas in the heartland of the Swiss chemical industry in Basle.

In Japan, opposition groups showed perhaps the best immediate response to Bhopal. Nineteen consumer, citizen, environmental and women's groups formed the Bhopal Disaster Monitoring Group. This Group both attacked UCC and drew attention to their own country's chemical industry. On 20 December 1984, group members demonstrated outside UCC's headquarters. Wearing black hoods decorated with skulls, they handed out leaflets pointing out that Bhopal is everyone's problem. The group wrote to seven Japanese chemical companies which produce MIC-based biocides, asking whether they had production plants in developing countries, whether any accidents had happened at these plants and what safety measures they enforced. The Group also met the Japanese Ministry of Labour's top safety authorities on 10 January 1985 to obtain detailed information on phosgene and MIC production, use and safety controls in Japan. However, the Japanese industry was able to allay public fears by pointing to heavy capital spending on health and safety. A Mitsui Toatsu Chemicals official said Japanese companies 'are spending a lot, maybe 10–20 percent of their production costs, to run safety and pollution abatement equipment'.

USA

It was in the USA that the chemical industry was forced to undertake a major campaign to respond to Bhopal. Under the Reagan regime, US toxic capital had scored major victories in strangling government regulations and in dismantling regulatory obstacles to its progress erected during previous regimes. Under Reagan, enforcement by the Occupational Safety and Health Administration (OSHA) had been cut back: charges of serious violations of safety and health standards were at half their 1980 levels. OSHA had abandoned inspections of companies with 'good safety records': this included many UCC plants.

Frank E. Mirer, health and safety director with the United Auto Workers, said: 'People should know that the Occupational Safety and Health Administration is not looking over anybody's shoulder at the current time'. Of course, even if they had been, it would not have made much difference. Margaret Seminario, health and safety specialist with the US AFL-CIO trade-union federation, was later quoted: 'Americans should not regard Bhopal as unrelated. The fact is that none of the conditions which led to the disaster would have been violations of specific standards or regulations of the Occupational Safety and Health Administration or the Environmental Protection Agency' (*Chemical Marketing Reporter*, 1 December 1986, p. 5).

Critics of the chemical industry used Bhopal to attack the lack of regulations to deal with a Bhopal-type situation. Indeed, so ill-prepared was the US regulatory apparatus for dealing with the issues Bhopal raised that the first job of the Environmental Protection Agency's Bhopal task force was 'to set up an internal information network so that people in the Agency's various offices will know how the Environmental Protection Agency is supposed to respond to a chemical incident'. The EPA, however, had no intention of being rushed into anything: over the weekend of 15 December 1984, it refused even to consider regulating MIC as a hazardous air pollutant.

The failure of the regulatory apparatus to respond adequately to the issues Bhopal raised left the running to the politicians. Congressman Henry Waxman (Democrat, California) said he would introduce legislation to force the EPA to regulate MIC. UCC's chief executive officer Warren Anderson pledged his assistance in drafting such regulations. Indeed, so important was it to UCC to be seen to be on good behaviour in the immediate aftermath of Bhopal that UCC became a firm supporter of increased government regulation of toxic chemicals. Anderson proceeded to wrong-foot himself with the rest of the chemical industry when, in his new-found zeal for increased regulation, he called for the adoption of national standards for hazardous air pollutants.

The original pressure for regulation in the USA centred on control and reporting of chemical emissions. This is a major sore point for the industry, as routine emissions of toxic chemicals are normal practice. This was shown in an EPA report on the Kanawha Valley, where UCC's Institute, West Virginia, plant is sited; it described the regular release into the air of large amounts of eight toxic chemicals, including known cancer-causing chemicals, as 'inherent to the large-scale production of any chemical product'. Attention was drawn to plant emissions by revelations of MIC releases at Institute.

Following up this question, Congressman Waxman wrote to major US chemical companies asking for details of emissions of specific poison gases. The request related to a list of twenty-four chemicals that are at least as hazardous as MIC; the list was prepared for Waxman by UCC. The responses by some of the companies gave some idea of the problem. Borg Warner replied that its Parkersburg, West Virginia, plant releases more than four tons per day of acrylonitrile, a cancer-causing chemical, and butadiene. Dow Chemical's Midland, Michigan, plant was reported to release more than six tons per day of methylene chloride, a suspected cancer-causing chemical, while an Exxon plant in Baton Rouge, Louisiana, annually emits more than 560,000 lb of benzene, known to cause leukaemia.

In February 1985 members of Congress, led by Representative James Florio of New Jersey, proposed a new round of regulation. This legislation included extensive mandatory right-to-know disclosure by chemical companies, limits on hazardous air pollutants and the development of emergency evacuation plans. While the legislation was prompted by Bhopal, Florio was at pains to point out its relevance to the US situation; he noted that there had been thirteen separate chemical spills and emissions in the previous three months in Linden, New Jersey, which sent hundreds of local people to hospital. Guy Molinari, a Republican from New York, who supported the Florio package, said: 'The less the public knows, the more industry can get away with. It has become obvious that industry will release information selectively and on its own timetable. This is unacceptable. The public has a right to know what hazards exist in their communities and the Florio package mandates that critical information be provided to the public.'

So strong was the pressure for regulation that the US Chemical Manufacturers' Association (CMA) speculated that the Reagan draft polity on hazard export, which had been lying dormant in the White House since August 1982, might actually be issued. It wasn't.

US industry responded to Bhopal and the increased demands for regulation by making reassuring noises about how safe the US chemical industry was. Dr Geraldine Cox of the CMA said: 'The emergency procedures that we have here are much more sophisticated and we practise them on a regular basis.' In a rare display of confidence in American workers, the industry implied that the quality and skills of the US workers are much higher than those of Third World workers. The industry argued that plant design was as safe as was humanly possible but that some accidents were inevitable, given that it was impossible totally to eliminate human error.[4] US trade unions supported the industry position: both UCC and the trade unions representing workers at UCC's Institute plant reassured a Congressional inquiry that no problem similar to Bhopal could occur at Institute. Such veiled racist explanations were largely accepted by the US public. US companies also supported UCC's attack on peripheral countries' regulation of multinational investment: this contributed to restricting the problem to the Third World.

Nevertheless, the industry realized that more than reassurance was needed. The industry's attempt to allay the fears raised by Bhopal was headed by Monsanto, one of the most 'progressive' companies in the US chemical industry. Monsanto had previous experience in this field. In 1984 it had been instrumental in forming Clean Sites Inc. in an attempt to allay concern over toxic waste in the USA.[5] Clean Sites' chairperson was Russell Train, former administrator of the EPA and chairperson of the World Wildlife Fund, USA. Clean Sites was formed to work out mediation rather than litigation processes for dealing with toxic-waste dumps. In this initiative, Monsanto had allied itself with right-wing environmental groups like the National Wildlife Federation, two-thirds of whose members had voted for Reagan in the previous election. In January 1985, Monsanto announced a voluntary initiative to inform the public around its plants of the various hazards the plants posed. Some indication of Monsanto's position can be found by examining its response to Congressman Waxman's request for information on releases of poison gases. Monsanto provided information only on eleven releases of regulated chemicals in 1984, those releases which it was legally required to report. As Waxman commented, given Monsanto's recent much-publicized move to provide information to communities near its plants, its response to his survey was meagre and regrettable (*Chemical and Engineering News*, 8 April 1985, p. 20).

The industry followed suit. According to the CMA's Stover (1985):

Acting through the Chemical Manufacturers' Association, the leadership of the industry quickly perceived that telling the positive story of the industry's past performance in safety would not be sufficient. It was clear that the public wanted more. One of the questions most often asked was: 'What will the chemical industry do differently as a result of Bhopal?'

Whatever the US public might have wanted, what it got was CAER, the Community Awareness and Emergency Response programme, which when spoken comes across as CARE, more typical industry double-speak. What it didn't get was implementation of the most recent technical recommendations on the provision of emergency-relief systems.

The industry's CAER programme was released in March 1985, a few days after UCC released its report on Bhopal, as an attempt to head off major legislation on the right to know. 'The CAER programme urges chemical plant managers to make publicly available the same information on chemical hazards that the industry has for years shared with employees.'[6] Furthermore, 'the CAER programme calls for [material safety data] sheets to be made public and for the industry to put those pertaining to high-hazard materials in the hands of local emergency response and medical officers so they will have accurate information on which to act'. This programme has received major support from the industry: 160 of the 180 members of the CMA named senior-level co-ordinators to make sure the CAER programme was effectively implemented. The industry was not doing this out of the goodness of its heart: the CMA admitted one of their objectives in the CAER programme was heading off additional regulations.

The industry was reasonably successful in this endeavour. Congressman Waxman introduced a bill in May 1985 to force the EPA to regulate chemical emissions. Both EPA and the industry opposed the bill, with the industry warning that plants would be forced to close if the bill was made law. Waxman failed even to get enough support in his own subcommittee to move the bill forward:

A subcommittee aide attributed opposition to the influences of the chemical industry as well as to insufficient public awareness of the air toxics issue and how it is being handled (or not handled) by Congress. Public pressure simply is not strong enough at this point to convince committee members to vote against industry interests and for public health, the aide said. (Environment, November 1985, p. 45)

In those areas where public pressure was strong enough, the industry intervened in other ways. New Jersey, which is a bastion of the chemical industry, introduced a Toxic Catastrophe Prevention Act in 1985, which was characterized by the CMA's Stover as an example of the most severe regulation to be introduced in the USA in 1985:

The bill identifies a list of acutely toxic chemicals and requires facilities which produce, use or store these chemicals to register with the state. Each plant must complete a safety audit and submit it to the state within six months. The state will review the audits and has the authority to order changes in plant designs and processes *to protect public safety and health. This legislation will become law in New Jersey before the end of 1985, and similar legislation is likely to be enacted in other states next year.* (Stover, 1985, p. 19, my emphasis)

Obviously such regulation had to be headed off – or if not headed off at the pass, then certainly corralled. In January 1986 New Jersey introduced a bill requiring companies to provide plans to prevent toxic chemical releases. The bill's success was due to what was described as a 'very positive, very constructive role' by the executive director of the Chemical Industry Council of New Jersey, the chemical industry's lobbying organization. Two major amendments allowed the chemical industry to support the bill. The bill's original provisions were for outside consultants to draw up safety plans. The industry's response? – 'Look, we're the experts.' Mr Bozarth [Chemical Industry Council, New Jersey] said, 'Why not look at what we're doing already and take it from there?' (*New York Times*, 9 January 1986, p. B2). This proposal would cut down expenses, as outside consultants are notoriously more demanding and critical than in-house staff. The other amendment allows the companies to appeal at every step of the process to an administrative law judge and from there to the civil courts. Because of this, the industry believes the New Jersey EPA will not close any plants, though the bill grants it that power.

On the right-to-know issue, industry had accepted the inevitability of regulations. The right-to-know movement was started by American trade unionists in the late 1970s, when they pressed the Occupational Safety and Health Administration for regulations that would give workers the right to toxicity information on the chemicals they handled at work. When the prospect of national regulations from OSHA fell victim to Reagan, Bush and the Office of Managment and Budget, trade unions and environmentalists pressed in the different states for local right-to-know laws. These campaigns had varying levels of success, with the result that, by 1985, twenty-two states had approved such laws.

In circumstances where industry is faced with a patchwork quilt of local regulations, it always favours national regulations. This is not simply to avoid the costs of varying bureaucratic requirements but also to dilute the more advanced legislation adopted by various states. New Jersey, for example, has a nasty habit of producing highly advanced regulation of the chemical industry. This is no doubt due to the major chemical industry presence in that state: some 25 percent of US chemical output is produced in New Jersey. This also means New Jersey is more aware of chemical hazards: there are an average of twenty accidental chemical releases in New Jersey each month, with five requiring some evacuation of residents. Thus, following the passage of the OSHA regulations, the chemical industry went to court to have the New Jersey right-to-know law overruled, as it surpassed those national regulations. A national standard would be more satisfactory for the industry, as it would claw back the advances made by the right-to-know movement in the states where it is strongest.

The industry thus succeeded in responding to the legislative and regulatory demands brought on by Bhopal in a way that decreased the threat to its continued operation on its own terms. Its strategy was to pre-empt government regulations by introducing its own programmes to satisfy the demands that produced the government regulations. By voluntarily providing information on the hazards that its operations posed, it prevented more stringent regulation.

While the industry presented itself as supporting the provision of information to the public around its plants, it was concerned that this provision of information be handled correctly. Robert Pyle, a spokesperson for UCC, told a US Senate committee: 'Community right-to-know programmes must focus on effective communication. Merely making information available to everyone who requests it may result in poor communication, raising needless apprehension, misinterpretation and unfounded public fear.' The industry's CAER programme made sure information on hazards was reassuringly packaged with information on the industry's plans to avert or control hazards.

Regardless of providing information on hazards to communities, the industry's stance on providing information to workers showed no major change. When Senators Metzenbaum and Stafford introduced legislation in March 1986 to ensure that workers exposed to hazardous chemicals at work receive proper notification of the risks that threaten them, the industry response remained true to form: hostile. The CMA claimed that industry, through current law and voluntary efforts, was already educating workers in this area. The right-to-know standard promulgated by OSHA and accepted by the industry has also been heavily criticized. Robert E. Wages, a vice-president of the Oil, Chemical and Atomic Workers' Union, told a Congress subcommittee that OSHA's hazard communication standard simply institutionalizes existing ineffectual company programmes. Wages argued that the standard should focus on labelling rather than on the provision of material-safety data sheets, leaflets detailing information on the effects and dangers of toxic chemicals produced by the companies that manufacture and use them. The problem with these material-safety data sheets, according to Michael Wright, of the United Steelworkers of America, is 'quality. It has been poor, and, for the most part, it still is.'

Similarly, Rodney Wolford, health and safety director of the International Brotherhood of Painters and Allied Trades, criticized the OSHA standard for failing to require evaluation of labelling of toxic chemicals. Because this evaluation is not required, the only certain outcome of the OSHA standard is a reduction of manufacturers' liability. As an example Wolford cited that while labels may warn painters not to breathe paint vapours, they do not say that neurotoxic symptoms such as headaches and dizziness indicate serious over-exposure. By allowing the companies to produce the information and by not evaluating it, OSHA was failing to protect workers. Companies will include only such information as they have to. Controversial or long-term effects of chemical exposure are sure to be ignored.

The chemical industry scored a major victory by maintaining control over what information is made available and in what form. The industry also planned ahead to influence the range and content of possible future government regulations. As one example, in May 1985 the American Institute of Chemical Engineers announced it would form a Center for Chemical Plant Safety. This Center was to examine acceptable industrial-safety practices and, by improving voluntary safety procedures, to act as a basis for future regulatory change. Though the Center was prompted by Bhopal, its aim is to improve plant safety in the US only.

6 INSTITUTE, LIKE BHOPAL

By the end of July 1985 it appeared as if the crisis presented by Bhopal had been successfully managed. While UCC had to increase capital spending on safety and found some of its restructuring in Europe blocked on safety grounds, its share prices had returned to their pre-Bhopal level. There was a weakness, that would be later fully exploited by GAF, in that many of its shares were now held by take-over speculators. UCC's restructuring had been given an added impetus by Bhopal: the company was going ahead with its plan to divest itself of surplus labour and to sell off certain of its businesses. Indeed, in terms of protection from the take-over mania that was then sweeping corporate America, Bhopal seemed to give UCC extra protection: chief executive officer Warren Anderson was of the opinion that the liability involved in Bhopal was a better poison pill than money could buy. By taking a tough line in the negotiating process, UCC had already brought down the generally accepted price of damages below the $1 billion mark and had barely shifted itself from its $200 million position. UCC's business in general had not been hurt, as other companies had shown solidarity by continuing to give the company orders. Nor had any major consumer boycott been launched against the company.

For toxic industry in general, the publicly visible crisis appeared to have passed. While increased regulations in the USA were inevitable, and right-to-know victories were won in the USA on the backs of the Bhopal victims, few other countries showed any inclination either to tighten or to increase regulations or deny toxic capital entry. Such international guidelines and regulations as were planned could easily be ignored and were likely to prove as powerless as previous attempts to regulate multinationals. Indeed, in dealing with peripheral countries, toxic capital now had a major argument against local dilution of multinational control. Various recuperative measures were under way in the USA to maintain public confidence in the industry. The one crisis the industry had not succeeded in managing was the insurance crisis, already started before Bhopal but given extra impetus by Bhopal. (It's harder to fool capital than the people.) This insurance crisis, though a major one, was not public. Regulation of the industry in the USA appeared manageable, with the prospect of some of the industry's own measures being accepted as a model for further regulation.

Now that the storm of media attention was well over, and perhaps because Anderson had not succeeded in settling Bhopal by his deadline of July 1985, UCC began to take a much harder line in negotiations with the Indian state. At this point UCC began to blame the Indians outright. Anderson said, 'We put the plant outside the city. They allowed the people to settle around it. The Third World has a lot to learn.' UCC now was talking of indefinite litigation, with Bud Holman saying, 'We'll litigate these cases one at a time if we have to.'

Company Town

Respite from public pressure was ended by a major leak from UCC's plant in Institute, West Virginia, in August 1985.

Why Institute? 'Our chemical problem originated because Institute is a black community. The chemical industry tends to put its plants in communities that are black or poor, or both. Institute is a company town. We're in the Third Word,' Estella Chandler, spokesperson for an Institute activist group called People Concerned about MIC, argues (*Charleston Gazette*, 19 August 1985, p. 5A). Until the 1960s UCC hired black people in Institute only as janitors. It vigorously opposed a proposal in 1982 to incorporate Institute and levy taxes on the UCC plant. Following the leak in August 1985, UCC began to talk to Institute citizens who wished to incorporate the town. 'We are exploring ways we might make more of a contribution to the Institute community', as UCC spokesperson Thad Epps put it (*Charleston Gazette*, 28 November 1985, p. 5A). UCC is the major employer in Institute. Similarly, it is virtually the only employer in nearby South Charleston.

Bhopal directed attention to the safety of the chemical industry, not only in Institute, but in the Kanawha Valley, where Institute is situated. This was long overdue. In 1978 the West Virginia Citizens' Action Group had published a report listing 225 different chemicals emitted from plants in the Kanawha Valley, ninety-four of which they rated as highly toxic. The West Virginia State Air Pollution Control Commission says that 300 chemicals are routinely being emitted. Eighty of these are significant in terms of their health impact, depending on the amount involved, according to the Commission. A study which claims that an increased risk of cancer in the Valley is related to chemical plant emissions is strongly contested by the industry: 'I'm not aware of any statistically valid study that indicates any relationship between the cancer rate and chemicals in the Kanawha Valley,' stressed UCC's Thad Epps (*Charleston Gazette*, 28 November 1985, p. 4A).

There were two important reactions to Bhopal in the Kanawha Valley. On 29 January 1985, a class action suit was filed against UCC by three residents of West Dunbar. This suit was filed after the publication of the EPA report on MIC leaks at Institute. A petition denouncing this suit was later signed by more than 1,500 people, including many local officials. As a recuperative move ('We have to study it before we do anything') a National Institute for Chemical Studies was set up in February 1985 by community and business leaders not directly involved in the chemical industry. The Institute's board consisted of seven business people, two college presidents, a lawyer, one environmentalist, one labour leader, one doctor, one hospital administrator and one social services advocate. The Institute's objectives were to examine environmental issues, bridge the gap between the chemical industry and the public and encourage the growth of the chemical industry.

Gerald E. Beller, a professor at West Virginia State College, describes the context:

Alongside these developments have come requests that local communities shoulder greater responsibility for their own safety. [UCC] Chairman Anderson argues that communities 'cannot absolve' themselves 'from the responsibility to help minimize the risks inherent in all industrial operations'. The National Institute for Chemical Studies proposes a 'dialogue . . . to develop the general public's capacity to operate as informed participants in the process of deciding what to do about managing chemical risks'.

In short we are all to share responsibility for what chemical companies might do to us. It is not immediately clear whether this constitutes an attempt at co-optation and a shifting of blame, or an honest desire to encourage a more involved and informed citizenry. (Charleston Gazette, 4 December 1985)

As some indication of which of these possibilities is involved with the National Institute for Chemical Studies, Beller observed that 'NICS has been successfully pressured to fund . . . "selected studies, films, and presentations on . . . the benefits associated with the presence of chemical companies" '.

While the chemical industry funded this Institute, the West Virginia Citizens' Action Group and the Air Pollution Control Commission wanted the chemical plants to install the best available technology to deal with air pollution. In July 1985, the Commission's director asked West Virginia chemical plant managers voluntarily to reduce air emissions. The West Virginia Manufacturers' Association responded negatively two days before the aldicarb oxime leak at Institute. 'It is our *belief* that emissions at present levels do not adversely affect public health.' The emphasis is added to underline the scientific basis of these reassurances, a practice continued in the following response from Thad Epps of UCC:

I think there's a feeling at Carbide that generally, and this does not include what happened Sunday, we have a strong belief we're not putting anything out there that is injurious to public health or the environment. On the other hand, we have to continue to try to do better to reduce emissions. What we're putting out today, we may find tomorrow to be a problem. (Charleston Gazette, 15 August 1985, p. 7A, emphasis added)

Following Bhopal, UCC's MIC plant at Institute was closed down. On 4

May 1985, UCC resumed production of MIC at Institute. A small protest was staged by some residents.

What Went Wrong

'Institute set us back somewhere near square one in terms of public opinion.'
– Laurence O'Neill, manager, environmental communications, Monsanto

'I considered Bhopal accidental and believed no chemical company would err like this in the US. The Institute release killed that belief.' – Anonymous UCC chemical engineer, South Charleston, West Virginia

Toxic capital's strategy of limiting its safety crisis to Bhopal and attempting to dismiss Bhopal as a rare, chance occurrence was undermined by the leak of toxic gas at UCC's Institute plant on 11 August 1985. The leak reopened the whole debate on the safety of toxic industry; it put to flight the easy racist answer that these things happen in India but can't happen here. It was impossible to dismiss the Institute leak in the same way that the Bhopal leak was dismissed. Indeed, the constant declarations that Institute was safe, and that nothing like Bhopal could happen there, intensified the crisis. Senator Byrd of West Virginia said: 'This incident has heightened my concern, as it has heightened the concern of others, because we felt that after the company spent about $5 million to put new equipment in place and emergency preparedness measures in place, that this could not happen.' Patrick Tyson of the Occupational Safety and Health Administration argued: 'The country was right in its expectations that this plant, of all the ones in the country, would be the one that was safe because of all the attention after Bhopal.'

It's worth quoting from the *New York Times* report on the leak (based on a UCC press conference on 23 August) to see how many things went wrong:

In the period leading up to the leak high-pressure alarms were repeatedly shut off and ignored. A high-temperature alarm was out of service. A level indicator in the tank that leaked and was known to be broken was not fixed. Meanwhile, the unit's computer, which silently recorded the rising problems for days, was never asked for the information by operators ... a total of thirty-two people in the sprawling plant had directed responsibility for the problem chemical unit in the days before the 11 August leak . . . But no one moved to correct the failures and the significance of those errors was not recognized until the toxic leak occurred.

Contrary to plant procedure, the workers never checked the tank and associated equipment before using it to make sure it was running properly – which it was not. The staff then did not programme its computer printer and terminals to display readings of the system's temperature, pressure and other conditions, so the operators did not know a major problem was developing. In addition, the workers assumed the tank was empty because a pump being used to drain it had stopped a few days earlier. But the workers never verified that assumption. Such a check would have been difficult in any case because the tank's level indicator was broken. (New York Times, 24 August 1985)

There were major similarities between the Bhopal and Institute disasters. Both were caused by runaway reactions. In both cases the safety systems were overwhelmed. UCC took hours to reveal what chemicals had leaked at Institute and played down their hazards. Information on the health effects of aldicarb oxime, one of the major constituents of the Institute leak was as lacking as information on MIC. UCC's failure to inform the local authorities promptly meant that the people of Institute learned of the toxic leak through the evidence of their own senses rather than through any advanced warning system. Both accidents resulted from multiple equipment failures, even though UCC would like to blame workers for the accidents. Both may be traced back to decisions based on capital's values of profit-taking and cost-cutting. Regulators and local authorities in both countries assured the population that both plants were safe, while actual inspection of the plants was considered unsatisfactory by the workers' trade unions (Jones, 1987).

Managing the Leak

'The things that have happened [toxic leaks at Institute and South Charleston] aren't any different than there have been for years. Bhopal just brought it out.' – Frank Leone, Mayor of Dunbar, West Virginia

UCC's management of the Institute leak closely resembled its management of Bhopal. Again we had the ritual displays of concern, the visit by the corporate president to the scene of the crime, followed by a later hardening of positions when the lawsuits came into play and the white heat of media attention had passed.

On 11 August when the leak occurred, UCC's medical officer Bipin Avashia reassured the population that aldicarb oxime is a 'very minor irritant' with 'no long-term effects'. UCC spokesperson Dick Henderson said 'the [early warning] system worked'. Attention was immediately focused on UCC's failure to warn local authorities in a timely fashion. Thus the first reaction centred not on what caused the leak but on the measures designed to deal with the leaks. UCC then admitted problems with this, but said it was an industry-wide problem. On 12 August UCC explained the leak had been caused by operating and design problems in a new process: this process had begun at Institute in response to Bhopal and was designed to avoid transport of MIC and thus increase safety. On the 13th, UCC admitted that the failure of the automatic warning system was due to the fact that their computer system was not programmed to deal with the chemical that leaked. UCC admitted this only after the computer system's makers, SES Inc., told the Press that the system worked perfectly but was not programmed for aldicarb oxime. UCC officials now also admitted that a problem existed with the size of pipes in the emergency-relief system but again pointed out that this was an industry-wide problem.

UCC suspended production of aldicarb at Institute until a special committee completed what UCC described as an 'independent' investigation of the leak. This committee was chaired by Russell Train, ex-administrator of the Environmental Protection Agency, chairperson of Clean Sites Inc. and a director of UCC. While UCC gave the impression at the time that this report would be made available, no such report was published and Train gave the report orally. On 23 August, UCC admitted that a major series of errors, both mechanical and by operators, led to the leak. Laurence Dunphy, assistant plant manager at Institute, echoed Warren Anderson's words at the launch the previous March of UCC's report on Bhopal: 'I can't know every single thing going on in every unit of our plant, but there are people in our organization who should.' (Institute has twenty chemical units and 1,500 workers.) Robert Oldford, president of UCC's agricultural products division, which was responsible for both the Bhopal and Institute plants, was replaced at his own request and placed on undefined 'special assignment' by UCC.

Some observers found it strange that UCC admitted so many failures in their account of the leak, since this exposed them to almost certain legal liability. However, UCC knew all this would come out eventually: it was better to have it all publicized in one fell swoop than to have it dragged out over months of government and Congressional investigations, with the attendant bad publicity. Since UCC was admitting the string of faults that led up to the leak (they were impossible to deny, anyway) its strategy was to minimize the effects of the release. Robert Kennedy, president of UCC's chemicals and plastics division and soon to be chief executive officer of UCC, said he was 'deeply and personally sorry for the fears and concerns' that the leak caused.

Jackson Browning of UCC insisted, however, that the released materials were 'by no means life-threatening' and would have no lasting effects. As usual UCC had no proof of this and had no intention of monitoring those exposed to see what the effects really were. It was left to the West Virginia health authorities to undertake a long-term study of those exposed. This study was undertaken when the authorities discovered that such animal tests as had been carried out on aldicarb oxime had exposed the unfortunate animals to a lower dose of aldicarb oxime than the even more unfortunate citizenry of Institute received.

UCC's reassurance campaign on the lack of toxicity of the leaked chemicals was backed up by the industry association. Geraldine Cox of the Chemical Manufacturers' Association told ABC's *Nightline* television programme that aldicarb oxime is less hazardous than aspirin (Everest, 1985, p. 54). On the question of aldicarb oxime's toxicity there were dissenting voices. The *Charleston Gazette* (14 August 1985, p. 1) quoted an unidentified independent health consultant as saying aldicarb oxime was seventy times more potent than MIC. UCC was backed up by independent government specialists, such as Donald Miller and Vernon Houk from the Centers for Disease Control in Atlanta, Georgia. Aldicarb oxime is 'far less toxic than MIC', Houk assured the public, though he added, 'I am not saying this is an innocuous chemical. It is not.' Once again these assurances were not based on actual testing. Houk and Miller said their estimate of aldicarb oxime's toxicity was based on aldicarb's toxicity, as little actual testing of aldicarb oxime had been performed.

UCC also received support from the local trade union and some local community officials. The Institute trade-union representative said he was pleased with the response to the leak. Joe Gresham, business manager for Local 656, International Association of Machinists and Aerospace Workers, said, 'We're going to have problems, but we're going to handle those problems.' Gresham pointed out the aldicarb oxime leak was stopped in fifteen minutes, while the Bhopal MIC leak continued for several hours. In a display of support for UCC, South Charleston city officials told Anderson that they would welcome the proposed UCC PCB-removal plant. This plant had been refused permission to set up in Paducah, Kentucky, where thousands of citizens had signed a petition against the plant. 'I saw where Paducah, Kentucky, said they didn't want the PCB-removal plant. I told Mr Anderson this morning we'll take it. I put my money where my mouth is. He said thanks for the show of support', Richie Robb, the mayor of South Charleston, told the *Charleston Gazette* in August 1985.

Robb was wasting his breath. Construction of the PCB-removal plant began in Henderson, Kentucky, in December 1985. Local officials had further occasion to be less than pleased with Anderson again later that year. While Anderson praised local officials during his visit in August 1985, the following month his story had changed: 'He told New York securities analysts that the valley didn't "have its act together" on evacuation plans' (*Charleston Gazette*, 28 November 1985, p. 5A).

This was part of UCC's general toughening up in what chemical-industry analysts described as the third phase in UCC's reaction to Bhopal. During this phase, which began after the gas leak at Institute, Anderson dismissed most of those affected by the Institute leak as hypochondriacs: 'I think if we had a leak of Arpège [at a plant] 135 would go to hospital.'[1] The UCC view now blamed the media for 'over-reaction' to both Bhopal and Institute. Anderson said he felt the suffering in Bhopal had been exaggerated: 'Today, the reports we get out of India . . . there's no blindness. They're only six to twelve cornea transplants. [It's] not as bad as was expected.' Thus Anderson reduced the effects of the worst-ever industrial disaster to six to twelve cornea transplants. This third phase in UCC's reaction, according to industry analysts, is 'a phase characterized by the company's insistence that its troubles are now in the past and that life has returned to normal. To many analysts, this is a delusion' (*Wall Street Journal Europe*, 29 November 1985).

Divided Community Response

The community response to UCC's management activities was divided. As at Bhopal, the area was visited by various UCC executives. On 15 August, Warren Anderson came to Institute for the opening of the National Institute for Chemical Studies, the private research organization sponsored by local business leaders from outside the chemical industry. On the morning of 16 August, being a keen amateur gardener himself (nice touch, this), Anderson held a press conference on the leak's effects on plants. At the press conference, Anderson set his argument within the cost–benefit line. Speaking at length on risk, he argued:

Someone has sold us a bill of goods that this is a zero-risk world. Life is not that way. I think in today's environment you couldn't invent the pencil. It has a very sharp point. Children use it. You can stick it in your eye or your ear. I doubt that you could get the pencil introduced into the market today. (New York Times, *17 August 1985, p. 6*)

In South Charleston, a march in support of UCC was organized. The organizers were three women who worked in a medical training office, but who had friends and relatives who worked for UCC. One of the organizers, Betty Ray, said the march was to balance the negative reaction to the recent UCC leaks. Ray said the 'silent majority' of Kanawha Valley residents were grateful to UCC for the thousands of jobs it provided. She added her own memories of the bad old days: 'I've lived here fifty years, and I remember when soot covered your feet and the stench was much worse.' As usual, the 'silent majority' failed to show up. Some 400 (or 600, depending on the source) residents endured light rain on the march to support UCC. Many marchers wore T-shirts and baseball caps labelled 'West Virginians support Carbide'. These were a donation from an anonymous South Charleston business person. Mayor Ritchie Robb addressed the rally, laying it on the line: 'This is a Carbide town. There's not a household in this town that doesn't have a member of the family or a relative working for Carbide' (*Sunday Gazette-Mail*, 18 August 1985).

UCC were less successful in Institute. A meeting organized by People Concerned about MIC, attended by 200–300 people, heard calls for UCC to pull out of Institute. UCC's Robert Kennedy responded to these calls in the time-honoured industry manner by comparing like with unlike: MIC is like tear gas; aldicarb oxime is like Arpège or a pencil.[2] Explain it simply to them, one can hear the consultants say. Put it in terms they'll understand. Kennedy's folksy approach began with a description of his childhood in the shadow of a Pittsburgh steel mill, 'where I had to turn on the headlight of my bike to go home at noon because of the smoke'. Kennedy admitted the existence of a great problem and expressed sorrow at the 'personal distress and emotional suffering' the leak had caused, thus reducing the leak's effects to psychological ones. Kennedy then argued that moving the plant from the area wouldn't solve the problem:

> '*I don't think we are an organization of quitters,*' *said Kennedy.* '*I don't think that we want to admit that we can't manage our own affairs.*'
>
> '*If we don't make those chemicals here, someone will*', *he said.*
>
> *To illustrate this point, Kennedy told a story about a dog he once owned that bit the mailman.*
>
> *Kennedy said that [he] considered giving the dog away, but a veterinarian told him,* '*Don't give that problem to somebody else.*' (Charleston Daily Mail, *19 August 1985, p. 5B*)

This is a superb example of chemical issue management. First, Kennedy reduced the exposure of people to a cloud of toxic gases to a psychological and emotional problem. Then he attempted to pass off an extremely hazardous chemical plant as a shaggy dog. According to one UCC engineer, the plant has 'enough toxic material . . . to poison the drinking water of every community from here to New Orleans' (*Datamation*, 1 March 1986, p. 41).

It didn't work. Kennedy's appeal was interrupted by a UCC worker called Eric Howard who shouted, 'I don't want to hear your dog story . . . You are manning these plants with untrained people. They are not trained – they are broken in by fellow operators.' After the meeting Howard told the *Charleston Gazette* about the leak the previous week in South

Charleston: 'There was no in-plant alarm Tuesday night. They were evacuating employees and nobody had turned on an alarm.' He also stressed the lack of safety in his own working area while unloading toxic chemicals from barges: 'I'm unloading toxic chemicals and the Coast Guard says you have to have two men. I'm unloading barges of toxic chemicals by myself.' On being asked what his view of loyalty to the company was, he replied, 'When I got hired at Carbide it was the best job in my life. But it ain't no good to me if I'm dead. There is a time for loyalty and there is a time for common sense.'

Kennedy told the Institute meeting, 'We are going to see how we can make these chemical plants the safest in the world.' Following the shop-floor description of safety at UCC, this can't have been too great a reassurance. No doubt his audience remembered similar claims after Bhopal. Given that it was unlikely that UCC could be got out of Institute because of its economic importance, citizens began to demand that they be provided with a way to get out of Institute should the UCC plant go out of control. On 29 September a march of residents protested at the lack of an evacuation route out of the town. On 19 November 1985, UCC announced the printing of 80,000 brochures telling Kanawha Valley residents what to do in an emergency. UCC was later reported (*Datamation*, 1 March, 1986, p. 42) to be 'supplying engineers, construction labourers, and $85,000 to build an emergency access road leading from the town of Institute'.

The overall response from the affected community was divided. This is only to be expected from people faced with the reality of living with toxic contradictions between health and work, wealth and safety. Since UCC is essential to the economic prosperity of the area, people have to believe that UCC isn't killing them softly. This means avoiding drawing certain conclusions and refusing to face certain questions. One person quoted in the local newspaper as a UCC supporter had two relatives who worked for the company who died of cancer. However she couldn't directly tie these deaths to their work; because of the economic support the UCC jobs provided her family with, she supported UCC. A constant refrain was how things used to be much worse and how the chemical spills were 'the smell of jobs'.

Regulation

There is little disagreement that regulatory inspections in India were inadequate, partly because of insufficiently trained regulatory staff: the two Bhopal inspectors were mechanical engineers with little knowledge of chemical hazards. Yet the situation in the USA was little better. Following the Bhopal killing, the Occupational Safety and Health Administration, the Environmental Protection Agency, the prestigious consulting firm A. D. Little and the local state authorities had investigated the Institute plant; they also inspected the safety improvements that UCC had made before the plant reopened. One UCC chemical engineer at Institute told *Chemical Engineering*:

One thing that amazed me during the review of the MIC production facility was the lack of technical knowledge, or lack of perception of how chemical units are designed and operated, that our government agencies show. They are perceived by the press as the watchdogs of our industry. Yet the West Virginia state agency employed two young engineers to review our technical information, calculations and installations of the unit. These chemical engineers were directly out of school (less than two years) and had no industrial experience. (9 December 1985, p. 27)

The EPA had concluded that safety there was 'above average' and said no federal enforcement action was currently planned against the company. The EPA was attacked by Congressman Waxman over this. Waxman said that the EPA was 'whitewashing' the threat to the public that UCC and other toxic plants pose: 'The Environmental Protection Agency seems to be giving assurances that no law has been broken, no harm is going to be done and that people should feel that they're being protected. What is frightening is that the Environmental Protection Agency has no idea what is coming from these plants yet blithely reassures the public that they are safe.' The use of regulations for reassurance was exposed by Waxman when he criticized the EPA's statement that Institute violated no federal clean-air standards: 'The Environmental Protection Agency didn't mention the fact that there are no standards because the Environmental Protection Agency hasn't set any. After fourteen years it has regulated only eight toxic pollutants.'

One major reason for the failure of the regulators was that their attention was narrowly focused on MIC. George Robinson, health and safety director with the International Association of Machinists and Aerospace Workers, said, 'We asked the government to be more rigorous in their inspections and to monitor the plant on a continuing basis. We felt the unit handling MIC was treated differently from other parts of the plant where safety checks were not carried out as they should have been by inspectors from the Department of Labor' (*Health & Safety at Work*, September 1985, p. 9). Patrick Tyson, the acting director of the Occupational Safety and Health Administration, attributed its failure at Institute to reliance on the 'traditional approach' of industrial hygienists: 'It probably wasn't the proper way to go' in inspecting Institute, he admitted. 'We may need to change our approach.'

In reaction to the leak, several teams of OSHA inspectors descended on Institute to do a 'wall-to-wall' inspection, expected to last six months. Immediate results of the inspection arrived in early October 1985, when UCC was fined for wilful neglect: three citations for wilful violations (price/maximum proposed penalty of $10,000 each) and three 'serious' violations (prices varying from $600 to $800). A spokesperson said UCC was cited for 'wilful neglect of numerous safety procedures', noting that OSHA rarely cited wilful violations – 'the definition of wilful is that they knew about a hazardous condition and did not make an effort to eliminate it'. Compared to this, the EPA remained, as always, a model of restraint: its report on Institute described UCC's response to the leak as 'relatively smooth'.

OSHA's full investigation led to 130 charges of 'wilful violations' in April 1986. Altogether $1.4 million in fines was proposed for a full total of 221 violations. Of the 'wilful violations', 129 were for failing officially to record work-related illnesses and injuries. Labor Secretary W. Brock, whose department is responsible for OSHA, told the US Congress in May 1986 that had UCC kept accurate records at Institute, Institute would have had 'an accident record substantially higher than the industry average' (*Chemical Marketing Reporter*, 12 May 1986, p. 7).

This charge has some interesting implications. Various chemical industry spokespeople have described UCC as among the safest US companies. On this basis alone it can be argued that very many more US chemical plants have worse accident records than the official figures. This means that the low injury average of the industry, the figure the industry normally uses to back up claims regarding its safety, is likely to be false. Secondly, it should be noted that in autumn 1984 UCC was fined small amounts by OSHA for failing to report asbestos-related lung diseases, yet these violations were described by the *New York Times* (5 December 1984) as 'minor'. Brock assured Congress that he wasn't making an example of UCC, saying similar health and safety violations had been discovered in other chemical plants. He reported that citations were being considered for such problems as failure to protect vessels from excessive pressure, absence of written safety procedures and absence of procedures to clean up cancer-causing chemicals which were regulated by OSHA.

UCC failed in an attempt to have the OSHA case dismissed before an administrative-law judge in Charleston, West Virginia, in August 1986. Some observers are cynical about the eventual outcome of the case. They note the precedent set in March 1986 when UCC and OSHA arranged a settlement relating to the August 1985 leak. In October 1985 OSHA had proposed fining UCC $32,000. The March 1986 settlement was for $4,400 for five 'serious' violations. An OSHA spokesperson said the settlement was agreed in return for a UCC agreement to correct violations immediately: 'Fighting the case could have taken two years and delayed the corrections, the official said' (*New York Times*, 20 March 1986, p. A23). As part of the settlement UCC agreed to purchase a $25,000 simulator 'that will enable workers at Institute to train on a chemical plant panel'. A similar agreement is expected in relation to the other proposed fines.

Warning Systems

Institute also showed how baseless are reassurances regarding the superiority of emergency warning and response systems in the metropolitan countries. The EPA report on Institute contended that 'local officials failed to follow standard emergency procedures, failed to set up emergency facilities outside the affected areas and failed to equip response crews properly' (*Chemical Marketing Reporter*, 23 December 1985). Other criticisms were made of the state response. Elderly and crippled residents of a nearby rehabilitation centre were not evacuated. The local authorities set up their emergency treatment centre at Shawnee Recreation Park: this was criticized by a former police officer who said, 'The park was engulfed by the gas and was the most vulnerable place around here' (*Charleston Gazette*, 19 August 1986, p. 5A).

The major problem relates to UCC, however. Following the Bhopal killing it was reported (*Sunday Times*, 9 December 1984) that every Institute resident receives every year a two-page letter detailing the various degrees of danger indicated by the UCC plant's various alarms. In the case of a full-scale leak, the police, sheriff and national guard were reported to have plans to evacuate the entire county. Yet few Institute residents said they had seen the letter UCC claimed to have sent every year since 1975.[3] Even if they had received it, they might have been no better off. According to Charles White, an official at West Virginia State College, 'If there are two blasts from the whistle that means a fire or emergency in the plant. If there are three blasts that means a gas release in the plant. If there are blasts every three seconds that means there's a danger for the people outside the plant. Now, I ask you, how many people are going to sit there with stop-watches and time the blasts?'

To continue with this example, if Institute residents hear a two-second blast every three seconds and see flashing blue strobe lights, it's a full-scale leak and they should be shit-scared. That is, if they can remember what the signal is and they didn't just throw out the UCC letter when it arrived. It also presumes they can hear the whistle.

In 1984, when a neighbourhood in Institute had to be evacuated at 3 a.m., following a valve breaking on a chemical barge moored at the UCC plant, most residents had their windows closed while sleeping and didn't hear the whistle. Even after the major overhaul of safety and review of emergency plans at Institute following Bhopal, this aspect hadn't improved: when in June 1985 UCC tested the whistle it would use in an emergency, just down the road at Andy's Grill customers didn't hear a thing (the *Wall Street Journal* reported). 'I guess I'm a goner,' Andy concluded.

The alleged superiority of public responses to chemical emergencies in metropolitan countries is well illustrated by the following account of an exercise at Carling, Moselle, France, on 21 September 1985. The exercise was to simulate what would happen if twenty tonnes of ammonia were to leak from the nearby Cdf. Chémie plant:

As agreed, early in the morning and without informing the inhabitants, the firm released a cloud of red smoke and sounded the alarms. Unfortunately, the response was limited. Although detailed instructions on what to do in the case of a dangerous leak had been circulated among the town's residents before the test, those who heard the alarms apparently did not consider themselves in real danger. As a result, most did not follow instructions to stay indoors and shut their windows. Instead, they went out to see what was going on and encountered about 400 observers stationed by the Protection Civile to count how many residents were still outdoors after fifteen minutes (by which time they would be reckoned dead). The volunteers, who weren't wearing gas masks, kept themselves busy asking the locals whether they had heard the siren. Still, a supermarket and three out of four schools followed instructions. (Chemical Engineering, *9 December 1985)*

Institute also illustrated the paradox that 'technology . . . ultimately undermines the very ends it starts out to accomplish' (Fulano, 1981, p. 8). Not only did the warning system installed at Institute fail to improve safety, it actually contributed to the accident. The installation of this computer system provides a case study in the failure of a technical fix. Ironically, a system installed to improve safety and early warning procedures contributed to the delay in warning the local authorities about the accident.

The failure of this system was a direct result of economic decisions made by UCC. The MIC plant at Institute was closed immediately after the Bhopal killing to maintain UCC's position that no differences in safety existed between the two plants. That is, since UCC claimed safety at both plants was identical, it followed that a similar incident could occur at Institute. Thus UCC was forced to close the Institute plant until it knew the cause of the Bhopal leak and took steps to ensure such a leak could not happen in Institute. Yet UCC needed to reopen the MIC unit at Institute by April 1985 if it was to avoid taking a further charge against earnings. Failure to open the unit by April would put UCC behind in supplying the chemical in time for its main seasonal market.

In its publicity leading up to the reopening, UCC emphasized it had spent some $5 million to 'make a safe plant safer'. One safety innovation it stressed was the installation of a Safer Emergency Systems 'plant watchman'. These systems are basically a microcomputer linked to metereological sensors that detect temperature and wind direction. It's also possible to add gas sensors that sound an alarm if a gas leaks and automatic telephone diallers that can inform local authorities that an emergency has occurred. The system can also have the characteristics of various pieces of equipment in the plant programmed in. The system takes out of human hands the calculations necessary to determine whether toxic gas clouds are likely to spread beyond a plant site, their direction, speed and dispersion rate.

Incorrect results from this much-publicized computer modelling system confirmed management's own estimation that the toxic-gas leak would not spread beyond the plant site. The system thus contributed to the delay in alerting local authorities to the Institute leak. The system's makers, SES Inc., charged that UCC's failure to provide critical information to programmers designing the system was partially responsible for the system's failure. When the Institute operator used the system, as aldicarb oxime was not one of the chemicals the system was programmed for, the operator chose MIC as the most similar chemical. The operator had little choice in this regard, as the system suffered from what was coyly called a data deficit: it was programmed only for three chemicals – MIC, phosgene and chlorine. Owing to the differences between aldicarb oxime and MIC, the computer provided incorrect results.

The cause of this specific failure was economic. Gary Gelinas, president of SES Inc., said the Institute system was a rush installation to enable UCC to reopen the MIC plant as soon as possible. In its contract with SES, UCC paid for the system to be programmed for only ten of the hundreds of chemicals it uses at Institute. When UCC reopened Institute, the system was programmed only for MIC, chlorine and phosgene. UCC delivered data on the other seven chemicals to SES only a few weeks before the August leak. Aldicarb oxime was not among these seven chemicals either. Gelinas said there was no limit to the number of chemicals the system could be programmed for. Each chemical added costs around $500. UCC spokesperson, Thad Epps, on being asked why aldicarb oxime was not programmed for, replied, 'We've got literally hundreds of chemicals here. We must make some sort of judgement.'

This represents a major problem for the industry. What chemicals should the computer include? All of them? Only the most toxic? How do they decide, given the general lack of information on the toxicity of intermediate chemicals? What about the possible new chemicals or combinations of chemicals that may be formed during runaway reactions? Even ignoring the industry's penny-pinching attitude to safety, the decisions here are objectively difficult, given the massive ignorance regarding the toxicity and possible interactions of chemicals.

Nevertheless, penny-pinching ruled the roost at UCC. It bought only a basic model, for $80,000. It didn't buy the gas sensors that can be directly linked to the computer. UCC has ground sensors for MIC but not for aldicarb oxime, at Institute. Nor did UCC purchase the attachment that automatically informs local authorities. Nor did it store any information on the tank in which aldicarb oxime was stored, nor the valves that burst in the computer modelling system . . .

There are other major problems with the use of this computer modelling system. Plume dispersion models have been criticized as possibly involving a margin of error of 50 percent. An authority on plume modelling, who is familiar with the Institute plant, told *Datamation:*

> *The main danger is believing the 'impressive' picture created on the colour graphics terminal without using judgement or looking out the window. Some officials believe that a plume modelling system permits you to evacuate fewer people in the event of a leak than you would otherwise. But that may not be the right move. When you consider that plume models have only been validated in* three basic tests *with ammonia, liquefied natural gas, and rocket propellant, a significant error factor could arise, especially in predicting the path of other materials.*[4] *(1 March 1986, p. 41, my emphasis)*

Safest Systems

'You can design the best system but when you deal with people you can create a problem.' – Geraldine Cox, technical director, US Chemical Manufacturers' Association (New York Times, *16 December 1984, section 3, p. 1)*
'We try to design safety into our systems' – Geraldine Cox

Unfortunately for toxic capital, Institute also showed how baseless were the industry's reassurances after Bhopal that chemical-plant design was the best and safest technically possible. It did this by bringing to public attention a major technological problem within the industry related to the design of emergency-relief systems, but also by showing the more general problems relating to plant design and worker response to systems malfunction. Following the Institute leak, chemical consultants told the *New York Times* that up to 50,000 chemical units in the USA were not designed to prevent hazardous materials from leaking if chemical processes got out of control.

The major design problems centred on the size of valves and pipes necessary to relieve pressure from tanks in which runaway reactions take place. The problem was already well known within the industry: as Ian Swift, chemical-engineering consultant, said, 'The technical lessons taught by the Bhopal tragedy are unfortunately not new ones.' Dissatisfaction over the design of emergency relief systems had led in 1976 to the formation of the Design Institute on Emergency Relief Systems. This Institute was a joint undertaking by twenty-nine US and foreign chemical companies under the direction of the American Institute of Chemical Engineers. This Institute was basically concerned with designing emergency-relief systems that would allow for two-phase flow (that is, they could accommodate not just vapour, but also liquid).[5]

It is a measure of the racism that permeates the whole Western reaction to Bhopal that the design of emergency-relief systems became a major issue only after the Institute leak. No doubt the issue was raised by the first major presentation of the Design Institute's findings, which coincidentally appeared in the July/August 1985 edition of *Chemical Engineering Progress*, the American Institute of Chemical Engineers' Journal. But the detailed findings had already been presented at the American Institute of Chemical Engineers' loss-prevention meeting in Houston in March 1985. They had also been previewed by Ian Swift in an article in the *Chemical Engineer* in August/September 1984. Indeed, as early as 1980 the group had concluded that emergency-relief systems two to ten times larger than those currently in use were required.

These findings were unquestionably available to UCC. Not only was UCC one of the sponsoring companies but Harold Fisher, a UCC engineer, had chaired the group in charge of this Institute's programme.

This confirms earlier comments by chemical engineers that UCC was aware of the problems regarding the Bhopal system but did nothing about them. One former UCC engineer said the failure of the scrubbing towers at Bhopal could have been predicted from their size, which would have been specified in the US design standards: 'The company is perfectly aware that these [runaway] reactions exist, but fifteen years ago when the plant was designed, safety designers didn't always take these things into consideration.' While the Bhopal plant was equipped for MIC in 1979, 'once a company has a working plant design, specifications are rarely altered' (*New Scientist*, 13 December 1984, p. 4). Ironically, one of the health and safety services UCC had been planning to market as part of its diversification programme was a procedure using a piece of apparatus to predict the size of scrubbing tower needed to contain a runaway reaction.

Speaking at the Chemical Industry After Bhopal Conference in London in November 1985, Ian Swift said: 'The results of the Design Institute on Emergency Relief Systems project indicate that relief systems using older techniques may be grossly undersized. This can be attributed not only to the increased requirements for two-phase flow but also to the possible, unsuspected, inadequacies of the runaway reaction rate data obtained under less than ideal conditions.' According to Swift, the failure of previous design methods was due to their being based on laboratory tests that seriously underestimated the temperature and pressure, owing to the scale of the experiments. (This is the old scale-up problem that has also plagued the nuclear industry. For this problem in the chemical industry see Davis, L. N., 1984, Chapter 10.)

One major problem in the study of emergency-relief systems is yet again lack of information:

The most serious disadvantage of using ... any ... systems modelling approach, in emergency relief systems design, is the lack of available data to use it effectively. One of the reasons why the styrene polymerization reaction was chosen as the large-scale test experimental system was that it was one of the few available for which complete kinetic thermophysical property data were known. In most cases, the runaway reaction products are seldom identified, let alone characterized. (Swift, 1985, p. 11)

This is exacerbated by the tendency of companies not to release information on accidents. 'Reports of actual incidents – when and if they ever become available – seldom contain sufficient information to allow a critical evaluation of the ERS [emergency-relief system] design adequacy.' Companies have refused to publish information on accidents in their plants despite the fact that such information could have been used to improve safety on an industry-wide basis. UCC refused, for example, to publish information on an accident in its Antwerp plant in Belgium in 1975 which killed six people (*New Scientist*, 10 January 1985, p. 3).

Opinions vary as to whether the industry has improved in this area. Consider the replies given to a question in the *Chemical Engineer* as to whether companies are forthright in sharing information among themselves:

Ventrone [editor, Plant/Operations Progress, *published by the American Institute of Chemical Engineers, and loss-prevention consultant] will give you a flat 'No' for an answer. He attributes much of this attitude to the introduction of a government agency, the Occupational Safety and Health Administration, along with corporate lawyers' control over information in cases of anticipated lawsuits. Brasie [associate process consultant, Dow Chemical] agrees that companies are not giving as much information as they might. The trend is for them to give only enough information to provide an*

explanation that is reasonable and plausible for the happening in question. Gearinger [division engineer, Du Pont] supports the thesis that the companies are not as open as they used to be; perhaps they never were quite as outgoing with information as they might have been so the industry could get the most benefit. (July/August 1985, p. 35)

Of course, a minority position disagrees: Grossel, manager of the process technology and design methods section, Hoffman LaRoche (who else?), told the *Chemical Engineer* he thinks companies are more forthright than they used to be.

The emergency-relief system design process suffers from other problems as well. The most difficult step is deciding what upset conditions they should be designed to deal with:

Although quantitative risk assessment techniques allow probabilities of different upset events to be estimated, their use in the chemical industry is limited by both practical, economic and philosophical issues. In practice, sufficient data are seldom available to assign accurate probabilities of various failure events, and the assumptions made will reflect the bias of the risk analyst or team involved... For complex processes many hundreds or thousands of man-hours may be required for the analysis, the cost of which must be borne by the anticipated profits of the process. (Swift, 1985, p. 4)

What all this means for the industry is trouble. Swift told the *New York Times* he had checked over one hundred emergency-relief systems in the past few years: all of them were inadequate, that is, too small to handle the flow of material resulting from a runaway reaction. He estimated $2 billion as the costs of fixing the industry's overall design problems in this area. The cost is one major reason why the industry hasn't implemented the Design Institute on Emergency Relief Systems' recommendations: 'The reason why two-phase flow techniques have not been widely used can be attributed to the lack of consensus over their applicability, their relative complexity in application and the cost of implementing their requirements' (Swift, p. 2). This lack of consensus means that, 'Although essential to process safety, no industry-wide standards exist for the design of emergency relief systems.'

Similarly, Swift told the *New York Times* (19 August 1985, p. A17):

'Many in the chemical industry are loath to retrofit, because the systems have run ten or fifteen years without problems.' But the fact that systems have run 'safely' is no guarantee that they will continue to do so under emergency conditions: 'We are talking about rare events here. The older the equipment, the more likely there will be problems if a runaway reaction occurs.' According to Swift, the age of much US chemical plant compounds the problem: old equipment is less precise and responsive than new computerized and automated systems.

Decisions on emergency-relief systems have to be made within the industry's cost limits.

The cost of a catastrophe like Bhopal cannot yet be determined. It is possible that it may far exceed the likely profits that might have been generated by the SEVIN process in India. The cost consequences of failure as well as its probability needs to be balanced against likely profits in order to realistically assess the viability of a particular system. Indeed the cost of upgrading the safety systems of existing plants, not adequately protected, may be too much of a burden for the industry to bear.

Swift summed up at the Chemical Industry After Bhopal Conference: 'The industry is now aware of the problems and has available the design techniques to solve them. Whether it has the inclination or resources to apply them remains to be seen.'

7 FALL-OUT FROM INSTITUTE

CHEMICAL MAN'S BURDEN

Whether 'mini-Bhopals', accidents such as that at Institute, are a normal part of the chemical industry's operation became a controversial question after the Institute leak. Some observers attempted to dismiss Institute as a UCC problem: Amantha Raman, a chemical-industry analyst, for example, said Institute 'defines Carbide's problem in running plants as global – it is not confined to any one location'. Unfortunately for toxic capital, it was a bad week for chemical 'incidents' in the USA: the Institute leak was accompanied by spills and other accidents at Charleston (West Virginia), Camden (New Jersey), Valentine (Arizona) and Fairfax (Virginia).

Hugh Kaufman of the EPA called the week in which the Institute leak occurred 'a typical week . . . The public is starting to recognize the magnitude of the problem engineers in the field have known about for years.' David Doyer of the environmental pressure group, the National Resources Defense Council, agreed that leaks like Institute occur all the time: he widened the criticism of the industry by noting, 'Literally billions of pounds [of chemicals] are let go into the air each year. It's a shame that the way to get attention to the problem is to have a disaster.' On the industry side, D. V. Gagliardi of Dow Chemical said, 'With what the public has seen this week, they might think accidents are routine. They are not.' But B. W. Carrh of Du Pont admitted, 'We are going to get a lot more scrutiny from the public and from Congress and it may be warranted.'

Kaufman is of course correct: if the company involved hadn't been UCC nor the plant Institute, it's likely the leak would have made only the back pages of the *New York Times*, as happened with twelve other toxic disasters earlier in 1985.[1] The normality of accidents like that at Institute was confirmed by the industry consultant, Om Kharabandan, who, addressing the Chemical Industry After Bhopal Conference in November 1985, held the view that 'mini-Bhopals' occur all the time. This view was repeated by other speakers at the conference.

While industry argued that such accidents are uncommon, the tide seemed to be turning against them. Both the EPA and the Attorney-General of New York State commissioned reports which showed such accidents were indeed common or normal features of industry operations. The Chemical Manufacturers' Association's Stover said Institute regenerated legislative momentum in the USA. As Senator Lautenberg (Democrat, New Jersey) said: 'After Bhopal, there was still some complacency in Congress. Institute is going to change things. The chemical industry's back is against the wall and there is going to be greater regulation.'

Industry critics responded to Institute with calls for more stringent regulation. Peg Seminario, associate director for occupational safety and health with the US trade-union federation, the AFL-CIO, said: 'When Bhopal happened, there was a general feeling in the media that it was a unique kind of accident in a faraway place . . . The factors we found were nothing particularly unique to the plant, or to India, and none of the things that led to the accident would have violated US safety and health standards.' Seminario argued that the regulatory response in the USA was

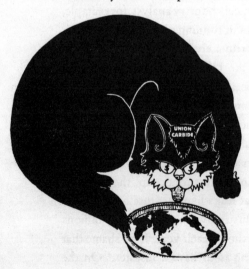

inadequate: 'When you look at all the regulatory activity, none of it gets at the hard issues of control . . . an adequate response just hasn't happened' (*Charleston Gazette*, 28 November 1985, p. 5A).

Critics now demanded adequate regulation. The National Resources Defense Council called for tighter regulation 'on the order of what the nuclear industry has had to face': this should include stringent standards on secondary containment systems, plant engineering and performance standards. This call for quality control of chemical-plant equipment was a new and threatening development for the industry. The need for this was underlined by A. Karim Ahmed, a US environmental-law specialist, who contended, 'People assume the chemical industry is high tech, but many plants use a nineteenth-century concept of plumbing when it comes to safety devices.'

Institute's legislative effects are undeniable. To quote Stover again: 'The Superfund bill that passed the Senate in September [1985] incorporated Right to Know/Emergency Response provisions essentially acceptable to the industry. It also included the objectionable emissions/discharge reporting requirements as well. In the House of Representatives, a similar emissions reporting provision was added to one of the pending Superfund bills.' The industry's oldest tactic – 'Where's the science?', more cynically described as 'Show me the corpses' – may not work in the post-Institute atmosphere.

Undoubtedly the industry will continue to argue that a sound scientific basis is prerequisite to undertaking significant new levels of emissions controls. But the industry also recognizes that the events at Bhopal and Institute have sharply advanced these issues on the Congressional timetable and may have made it more difficult for some members of the US Congress to accept the 'good science' argument. (Stover, 1985, p. 32)

Stover describes the context in which this regulation is pending:

Health and environmental measures are the latest of four major factors causing fundamental restructuring of the industry. The others are maturing products, international competition and changes in energy matters. A number of chemical plants in the USA are shut down, some permanently. Should the Bhopal aftermath trigger a series of new and costly regulatory requirements, heavy costs and outright restrictions would be the certain result and competitive positions would be affected. There is considerable doubt that prices can be raised to recover these new costs.[2]

Stover went on to warn that increased regulation might force US companies to build new plants overseas.

Bhopal and its fall-out came at a bad time for the US chemical industry. The industry was beginning to recover, with 1984 sales of $211 billion setting a new record, owing to increased production, not higher prices. According to Stover, 'Net income after taxes of $13 billion in 1984 equalled the all-time high of 1981, showing a satisfactory recovery from the depressed earnings of 1982 and 1983.' This recovery continued through 1986. Chemical companies' profits in the USA increased 54 percent, up to $13 billion, on sales that increased only 1 percent, up to $216 billion. Average return on equity climbed to 13 percent, up from 9 percent in 1983, excluding UCC (*Business Week*, 12 January 1987, p. 52). While production levels had increased, the industry had divested itself of much labour: employment in the chemical industry in 1983 and 1984 was the lowest since 1976, with only 1,060,000 working in the US industry.

Like the nuclear-power industry, the chemical industry can be squeezed by costs arising from increasing safety demands, though industry figures on the costs of regulation should always be taken with a large dose of salt.[3] Increased safety demands can drive up costs at a time when the industry is not producing a spectacular return on capital. UCC provides a prime example of this: despite its massive size, its return on capital had dropped to 4 percent, with a five-year average of 11 percent, even before Bhopal hit it. Obviously, the economic crisis is unevenly distributed among firms: those depending heavily on commodities are more affected than those relying on low-volume, high-value speciality chemicals. UCC is a poor example as well: one cruel description of it as 'the turnaround company that's always turning round' is not totally off the mark. Response to the crisis has varied, with some companies making major moves into biotechnology and pharmaceuticals. Nevertheless, increasing costs will squeeze the whole industry.

The major forces determining these costs are political pressure and the public perception of the industry. The level of safety demands on the industry is a reflection of the strength of opposition to toxic industry. Bhopal has had most effect in the USA because of the high level of opposition to toxic industry there. The industry has recognized the threat its public image poses: as Eric Draper of the US National Campaign Against Toxic Hazards has said, 'The effect of the tragedy in Bhopal is that the chemical industry has realized it has a public-relations problem. It has not recognized a public-health problem.'

The industry has recognized its public-relations problem and has undertaken sterling efforts to allay concern over toxic chemicals. Despite this, it is losing the major battle over the public's perception of the health effects of toxic chemicals.[4] Speaking at a Chemical Manufacturers' Association conference in Washington in October 1985, L. M. Thomas, EPA administrator, said Superfund and Bhopal had given the US people 'a very distorted view of the problems associated with toxic chemicals'. Citing a recent Roper poll, Thomas reported: 'Some of the most distressing findings deal with the public's perceptions of health threats, and their confidence in institutions such as government and private industry to deal with them. For one thing, people tend to attribute health problems more to environmental contamination from chemical wastes and pollution than to personal habits such as smoking, drinking or eating.'

Among other disquieting results Thomas cited were that nearly half of all US adults think the chemical industry isn't regulated enough; more than half express doubts about chemical-plant safety; and more than half believe the chemical industry has too much influence over government policy. Thomas's conclusions? 'Clearly, this is a major public-perception problem.' Thus a public-health problem is redefined as a public-perception problem. Thomas's solution? Rather than clean up its act, the US chemical industry's greatest environmental challenge is to 'build within the American mindset a belief that you care about their safety and their environment'.

The industry has recognized the depth of the problem. In a revealing interview, L. Fernandez, chairperson of Monsanto, described the chemical industry's crisis as part of a general crisis regarding the risks and benefits of science and technology. Fernandez described the industry's major task as overcoming the public perception that the industry puts profits ahead of safety and responsible operation:

Question: *Isn't it a logical perception when you seem to get one disaster after another from Love Canal to Times Beach to Bhopal to Institute, with assorted smaller spills and disasters along the way?*
Fernandez: *It's perfectly understandable. But we have not yet, in my opinion, as an industry put the amount of time, talent and attention – and financial resources – into the development of a communications and education programme to the public. We're an order of magnitude away from where we are going to have to be. And I think that we're going to have to have a long-range educational programme, not just a quick-fix communications programme. That is part of an even broader subject, and that is how to make the American public more comfortable with technology. I think you're going to have to go into the grammar schools. That means really long-range planning and thinking. It isn't just the chemical industry that is the problem, although we have our own particular piece of it. And we don't have to wait for the rest of the world to join us to be doing substantially more than we are doing . . . That, to me, is the biggest thing the chemical industry needs to be doing – addressing the whole issue of education and communication. (Chemical and Engineering News, 7 October 1985, p. 15)*

The Chemical Manufacturers' Association is again following Monsanto's line in this matter. Speaking on 5 June 1986, George J. Sella, Jr, chairperson of the Chemical Manufacturers' Association board, described what he called a 'gap' between public perceptions of chemical hazards and the action needed to deal with them and the industry's position. Sella believes that the gap cannot be closed, but that it can be narrowed. In a heartfelt cry, Sella continued:

Essential to the narrowing is communications. How can we make the public understand:
– the need to avoid impractical approaches?
– the need for cost-effective solutions?
– the need to follow good technical and scientific practices?
The Burden is on our industry.
– We must try to educate.
– We must take the initiative. (Sella, 1986, pp. 5–6)

The industry is moving to take up the Chemical Man's Burden. The major weapon here is the Community Awareness and Emergency Response programme (CAER). By June 1986, 180 companies, 90 percent of US productive capacity for industrial chemicals, had named CAER executives. Fully 1,500 plants are involved in the CAER programme and 900 of them have already 'sponsored the development of community co-ordinating teams', according to Sella.

The Burden does not stop there: 'Chemical companies and local CAER groups also are sending their technical personnel to talk in schools, churches and other community organizations.' R. W. Piercy, assistant manager of Olin Chemical Group's Joliet, Illinois, plant, explained that the goal 'is to demystify the chemical industry'. He says that the 'companies want residents to understand that the plants near them have responsible safety and emergency plans in co-ordination with their local officials' (*Chemical Engineering*, 10 November 1986, pp. 23–5).

E. Keith Young, associate manager of safety and industrial hygiene at CITGO Petroleum Corp.'s Lake Charles, Louisiana, refinery, describes the change in management towards the implementation of an 'intense communication programme':

For years, it has been standard procedure to seclude ourselves behind the plant gates and carry on the activities of plant management with little contact with the community in general. However, the recent incidents in Bhopal, Institute, West Virginia, and others have quickly changed the public's perception towards our facilities from benevolent employers to menacing giants. We can no longer carry on business as usual. We must become more aggressive in combating these erroneous perceptions. (Chemical Engineering Progress, *July 1986, p. 14)*

Risk Re-education

Much emphasis is being placed on the development of evacuation and emergency plans. The development of evacuation plans has been a key demand of the anti-nuclear as well as the anti-toxic movement. Some 370 local CAER co-ordinating groups have developed formal emergency-response plans.

CMA photo of school evacuation drill: would this measure protect the children in a real emergency?

Now, entire communities routinely practise what to do during emergencies. Hank Karawan, director of community awareness and emergency response at Union Carbide Corp.'s South Charleston, West Virginia, plant, and a member of the Kanawha Valley Emergency Planning Council, says that nine or ten such drills are being conducted this year throughout the Kanawha Valley – about twice as many as were being held a few years ago. (Chemical Engineering, *10 November 1986, p. 23)*

The development of these plans is a double-edged victory. On the one hand, they force the industry to admit that their operations are hazardous enough to require such evacuation planning; on the other hand, they allow the imposition of industrial discipline on the general public. Instead of chemical accidents being forbidden, and something to be prevented, we now have the industry training the public in how to deal with chemical accidents. Chemical accidents that affect the general public thus come to be considered normal or acceptable, requiring management rather than prevention. Similarly, the evacuation plans and practices become a way in

Re-educating for risk: CMA photo of emergency-response drill held at a Ciba-Geigy plant by the Geismar Area Mutual Aid Association.

which the industry reassures the public that, even if an accident happens, they will be safe.

The chemical industry resembles the Jesuits, not only in its desire to reach very young people, by educating them even in primary schools, but also in its desire to make friends and influence people in high places. Sella explained why he thinks the Chemical Manufacturers' Association has become 'one of the most effective lobbying organizations in Washington' and 'why our points of view and recommendations are being taken into account more and more': 'We have made a commitment and dedicated resources to providing well thought out and yet timely input to those crafting legislative and regulatory solutions' (Sella, p. 1). Thus the industry is continuing with its strategy of presenting a package for regulation based on its own voluntary activities: even if the industry's package isn't taken on wholesale by the legislators and regulators, it has a major influence on the permissible range of debate and regulation.

This closeness to the regulatory and legislative bureaucracy allows the industry to respond to future regulation and, if possible, to head it off by voluntary industry action. While the industry had been successful in beating Congressman Waxman's air-emissions legislation, the various states developed their own policies. Sella (p. 4) explains: 'Some states are moving toward percentage reductions overall . . . This certainly left the door open for us to come in with our own voluntary initiatives on controlling air toxics, and that is precisely what we have done.' Faced with possible regulation in this area, the Chemical Manufacturers' Association announced in March 1986 an air-emissions control policy, a 'voluntary, industry-wide effort to further assess and control plant emissions'.

The programme involves chemical companies in producing an inventory of chemical emissions, research into the effects of emissions and an analysis of existing emissions-control technology, reduction of emissions as considered necessary by the industry and communicating its actions to the government and local communities. Some reduction in emissions has already been announced. *European Chemical News* reported (31 March 1986, p. 13) that four companies had agreed to reduce chemical emissions in West Virginia in an attempt to forestall government regulations. UCC intended to cut emissions at Institute by 34 percent and at South Charleston by 42.5 percent; Diamond Shamrock was to cut emissions by 16 percent at Belle, West Virginia; Du Pont was to cut emissions at Belle by 19 percent and Monsanto was to cut emissions at Nitro by 23 percent. CITGO Petroleum Corp.'s E. K. Young reported how this works at the local level:

> . . . *the city of Lake Charles recently decided to establish a set of guidelines, in the form of ordinances, that would outline how emergency responses to spills, leaks, etc., would be handled . . . Our company has been volunteering to help serve on the Committee by sending Safety Managers. The city was happy to receive the expertise gained through many years of emergency responses. Our company's input into the ordinance making made it possible to exclude unnecessary ordinances by differentiating real risks from perceived risks and fears.* (Chemical Engineering Progress, *July 1986, p. 16)*

The industry is fortunate that its belief in the need for 'risk education' is not confined to its member companies. That the need for a propaganda campaign of risk education is also realized by the US regulatory authorities was shown by a revealing speech made by Lee M. Thomas, EPA administrator, to the National Conference on Risk Communication in January 1986. It is obvious from Thomas's speech that the risk assessors have simply not been successful in convincing the public that they are over-preoccupied with the risks of toxic chemicals and hazardous waste dumps, while they should rationally be more concerned about the risks of driving.

Therefore Thomas prescribes a course of risk re-education for the public. To educate the public, state and local bureaucrats must be made more conversant with how to inform people of the *true* nature of risk, rather than the emotional, misguided one the public currently holds: 'We must do what we can to increase the number of people who can communicate effectively about risk. State and local leaders must become more familiar with the language and skills of risk analysis' (Thomas, 1986, p. 7). He complains of the 'emotional arena' in which risk communication takes place. It is no wonder this is an emotional arena when Thomas can state: 'It is also *in the nature of things* that some places are riskier to life than other places and that some people bear more risk than others' (pp. 4–5, emphasis added).

More interesting yet is the type of risk that the regulators should be concentrating on: 'We have been accumulating evidence that in many places the major sources of health risk are not industrial plants or even hazardous waste facilities. They come from things like radon, a natural radioactive product of certain types of rock, from the air in homes, from wood-stoves, from the gas station and the dry cleaners down the street' (Thomas, p. 8). One can easily picture eager EPA risk communicators earnestly assuring the populations of such high-risk areas as New Jersey's 'cancer alley', or the Kanawha Valley, or Baton Rouge, that it's the wood-stoves, the filling-stations and the dry-cleaners, even the very rocks themselves, that they should worry about rather than those piffling petrochemical complexes down the road.[5]

Insurance Crisis

'Neither a business nor its insurer can make reasonable operating or financial decisions when liability is a wild card.' – Warren Anderson, UCC (Chemical Marketing Reporter, *17 November 1986, p. 70*)

Bhopal represents a major crisis for toxic capital, a safety crisis that can have a major effect on toxic industry's restructuring, though this crisis has mainly been confined to the USA. To some extent, the emphasis on Bhopal has skewed our image of this crisis, as the safety crisis existed at a lower level prior to Bhopal. The general safety crisis is accompanied and confirmed by a specific and highly expensive insurance crisis. This insurance crisis was already under way before Bhopal but was intensified by the safety crisis Bhopal produced, as well as by the uncertainty over the full extent of UCC's liability in Bhopal. Before Bhopal, the environmental-impact liability insurance market had practically evaporated. The sudden and accidental pollution insurance market had reduced individual companies' coverage from $300 million to $50 million, with even this lower figure unavailable to large chemical companies.

There are two explanations for the current insurance crisis. The first is that the crisis was caused by increasing court awards, including punitive damages, won against toxic capital in the US courts. The number of multi-million-dollar verdicts rose from one in 1961 to 251 in 1982 and 401 in 1984. For an example of how expensive toxic settlements can be, the total demands in suits filed for poisoning by toxic waste at Love Canal came to $1 billion, though obviously this was not awarded. This crisis could then be described as the return of the externality to haunt toxic capital's balance-sheet. Pollution claims 'hold out the prospect of widespread insolvencies among major carriers' according to the vice-president of Crum & Foster Insurances Companies. If US courts gave large US-style damages to the Bhopal victims, 'it could blow the lid off the international insurance market', Peter B. Bickett of Johnson & Higgins, a leading New York insurance broker, predicted.

The other explanation denies there has been a major increase in liability awards. It claims instead that the current crisis was caused by massive underpricing and cut-pricing by the insurance industry in a period of high interest rates. The industry was able to sell policies at low prices owing to the high interest rates the insurance companies earned on their premium income. The fall in interest rates undermined the temporary profitability this strategy gave the insurance industry. The industry then attempted to attribute this crisis in profitability to a litigation crisis, rather than admit it resulted from the industry's near-suicidal price-cutting.

There is a large measure of truth in the latter explanation regarding the general insurance market. This view expects increases in prices to remove the industry from crisis. In general, this seems correct. In the areas of product, environmental impact and catastrophe liability, however, the crisis is still continuing some two years after it became a matter for public debate. Insurance and reinsurance for these high-risk lines is rare, while the general insurance market has recovered by increasing prices. Some of these liabilities, in particular environmental and disaster liabilities, have been literally priced out of the market.

The crisis in these areas was characterized by huge increases in premiums. According to Larry Brock of insurance brokers Marsh & McLennan, 'A Fortune 1,000 company whose liability premiums cost $20 million last year [1984] would have paid $18.5 million against predictable small claims under $1 million. The $1.5 million would have bought it some $250 million worth of excess cover. This year [1985] it will be paying $4.5 million – three times as much.' In 1986, catastrophe-liability premiums were expected to increase three- to fourfold for large corporations.

As premiums increased, cover was lowered and some companies found they were unable to obtain insurance at any price. 'Our premiums for product and general liability and workers' compensation have gone up some 200 percent this year [1985],' according to asbestos producer Johns Manville's risk manager. 'But it's not only high costs that are the problem – trying to find cover at all is increasingly difficult.' To quote another insurance industry source: 'A lot of good industrial companies used easily to get $200 million to $300 million in liability coverage. Now they're lucky to get $50 million, and even then we want more money up front in premiums.' One trade group estimated demand would exceed supply by $62 billion until 1987 and, even at present prices, the industry's premium income in 1987 would fall short by 18 percent of the amount of business it could obtain.

By February 1985, only one major carrier, American International Group, remained in the environmental-impact liability market. In March 1985, the *Wall Street Journal Europe* reported that the pollution-liability field had almost collapsed, with all but three or four of the fourteen US companies and pools that used to handle the market pulling out entirely. The reduction in product liability was due to reinsurers abandoning the market, with the 120 companies writing it in September 1984 winnowed down to fifty or sixty by December 1985, with little change one year later.

Some companies have felt the brunt more than others. Acmat Corporation, one of the largest US specialists in asbestos-stripping, has been refused renewal of its insurance. As a result Acmat will have to form its own offshore insurance company or abandon the asbestos-removal industry. Biesterfeld US Inc., a subsidiary of a West German firm, abandoned the US market when its product-liability premiums were increased twenty-five-fold while its coverage decreased threefold. UCC was especially affected by the insurance crisis. In November 1985, industry sources were reported to be speculating that UCC might find its policies impossible to renew. Similarly, and of relevance to the behaviour of UCC management, speculation abounded as to whether UCC management had succeeded in obtaining directors' and officers' coverage. According to a UCC proxy statement, UCC's directors' and officers' coverage was cancelled after Bhopal. UCC and several directors have refused to comment on whether they are insured for events after 19 February 1985. 'If company directors and officials are without insurance, they could be personally liable in the event of lawsuits – meaning that if a suit against them is successful, they could lose their homes, cars and bank accounts,' the *Wall Street Journal* observed.

Other companies have had no alternative but to set up their own insurance companies. Toxic-waste disposal companies have found that no one wants to provide them with environmental-impact liability insurance. In response they are proposing to set up Waste Insurance Liability Ltd, a captive insurance firm. Another group of major US corporations – thirty-two giants including Du Pont, IBM, GE and Ford – has set up the American Casualty Excess Insurance Co. The guiding light behind this company is Marsh & McLennan, the world's biggest insurance broker, who persuaded some of their clients to set the company up when it was unable to find enough insurance for them. The thirty-two companies paid in $10 million each to set up the company. Premiums range from $100,000 to $2.5 million. It also offers directors' and officers' cover of $50 million. But even this company's coverage is limited: it won't cover aeroplane or nuclear insurance; its top pay-out for product-liability insurance is $100 million. (By way of comparison, A. H. Robins paid out $520 million in damages relating to the Dalkon Shield intra-uterine contraceptive device before it filed for protection under Chapter 11 of the US bankruptcy code.)

There are still other snags for the high-risk companies involved. The American Casualty Excess Insurance Co. will only offer directors' and officers' insurance if the corporation involved already has $25 million in such cover from other insurers. Furthermore, it will only pay out after the companies involved have exhausted their ordinary cover. Even with these strategies, many companies are underinsured. Both Ford and American Motors, for example, have reported in 1986 that they are fundamentally underinsured.

The insurance crisis is not confined to the USA, though it is most extreme there. Following the Sandoz leak in November 1986 in Basle, Switzerland, the European Council of Chemical Manufacturers' Federations was reported to be hurriedly putting together an insurance scheme 'to break the industry out of its increasingly claustrophobic liability insurance arrangements' (*Chemistry and Industry*, 19 January 1987, p. 38). The European scheme differs in its strategy from the American ones. Instead of directly insuring its own members, the industry is planning to set up a reinsurance group. This reinsurance group will provide back-up to the direct insurers of European chemical companies, who will be able to lay off some of their risk to the chemical-industry reinsurance group.

Shaping up behind this current insurance crisis is the spectre of an even more dangerous crisis. This coming crisis has two aspects: one relates to the insurance industry's ability to provide sufficient capital reserves to deal with catastrophes. The other relates to whether certain products or processes are or will become uninsurable. In a recent survey of the reinsurance market, the *Financial Times* (8 September 1986, p. v111) observed: 'Reinsurers face a bigger problem still, if not their worst problem: the provision of catastrophe reserves. As the world becomes more industrialized, the potential for catastrophic losses increases. Reinsurers have also woken up to the fact that, just because an earthquake or hurricane has not occurred in a particular part of the world, that doesn't mean it never will.'

The implication here is that the insurance solution to toxic catastrophes is limited. Lagadec has quoted the Munich Reinsurance Company as raising the question of the limits of insurance:

> ... *never before in the history of mankind have so many individuals lived so close to each other; never before have objects of such high value been found concentrated in so many limited spaces and never before have so many people had accidents simultaneously and of such severity and such frequency. For to the natural causes of accidents and disasters have been added those which man himself [sic] has created, causes which must be attributed to techniques and civilization.*
>
> *The existence of insurance companies will not change much in this respect. The foresight, the preventive measures against damages are only too often caught up with or surpassed by the appearance of still greater perils, by still greater concentrations of values ...*
>
> *The institution of insurance results from human reason. In large measure it permits the material reparation of the consequences of human failure. But it will also logically find its limits at the time when mankind no longer has the capacity to settle the problems of its existence in a reasonable manner. (1982, pp. 312–13)*

The Munich Reinsurance Company is still playing the role of the pessimist. The same *Financial Times* survey reports: 'It has also questioned whether something is insurable or not in the present seller's market. If reinsurers are not willing to cope with a risk, it is uninsurable – a strategy determined not only by price, but by the product itself.' The most obvious immediate example here is intra-uterine contraceptive devices. Continued lawsuits from women claiming to be injured by the Dalkon Shield have forced A. H. Robins into Chapter 11 bankruptcy. The beginning of a series of suits in the USA against Searle's Copper-7 intra-uterine device led to Searle's pulling the Copper-7 off the US and British market. The introduction of further birth-control products into the US market is likely to be stymied by massive insurance-liability difficulties.[6]

Similar questions exist regarding the insurability of biotechnological products and processes. As Richard Haayen, president of Allstate Insurance, told *Newsweek* (12 May 1986), 'If the risk is unknowable, then it is inherently uninsurable.'

> ## TOUGH TIMES FOR UNION CARBIDE
>
> **Autumn 1984:** UCC fined $50,000 for burning toxic waste in violation of state codes in West Virginia. The Occupational Safety and Health Administration also fined the company for what the *New York Times* (5 December 1984) described as 'minor violations such as failure to log cases of asbestos-related lung diseases'.
>
> **December 1984:** UCC fined $55,000 by West Virginia authorities for failing to report it was producing several hazardous chemicals at its South Charleston research centre. UCC described these as a 'bookkeeping error' and reported them before the Bhopal killing.
>
> — Killing at Bhopal.
>
> **23 January 1985:** The US Environmental Protection Agency said MIC leaked twenty-eight times from 1980 to 1984 at UCC's Institute, West Virginia, plant owing to 'equipment failures' and 'human errors'. An investigation found no evidence of injury from the leaks. The largest release took place on 1 January 1981: a broken feed-line released a 14,000 lb mixture of chemicals, including 840 lb of MIC. UCC said the mixture was collected and 'returned to scrubbers for clean-up'. Of the releases, seven were greater than 10 lb, seven were from 1 to 10 lb and twelve were less than 1 lb. UCC also spilled toluene, a solvent which causes neurological disorders, into the Kanawha River in December 1984 and January 1985. UCC failed to notify the EPA promptly of the December 1984 spill, as required by law.
>
> **January 1985:** Four hundred people were temporarily evacuated from UCC's Taft, Louisiana, plant at the end of January 1985 after a sharp rise in temperature in a paracetic acid and ethyl acetate storage plant. The same week, an explosion and fire were reported in the plant's ethylene unit. No chemicals were released, UCC said, and the fire was put out in thirty minutes.
>
> **30 January 1985:** UCC reported there had been thirty-three more

leaks of MIC since 1 January 1980 at Institute than those the EPA announced. These were all under 1 lb and were not required to be reported. Also since 1 January 1980, there had been 107 in-plant leaks of phosgene and twenty-two leaks of phosgene/MIC mixtures, but none had escaped the plant in sufficient quantities to break Federal regulations. Representative Joseph Gaydos said he was 'more than just a little angry' that the series of spills at Institute had come to light after December Congressional hearings at which UCC and union officials 'told us everything was fine and dandy at the plant'.

February 1985: UCC hit by controversy in Australia over the storage of 160 metric tons of dioxin-contaminated waste at its Rhodes chemical plant in Sydney, Australia, which had been left over from discontinued 2,4,5–T operations.

March 1985: Flemish regional authorities refused UCC permits to store 70,000 tons of chemicals at its Malle production site, saying studies showed the risks of such storage to be unacceptable. In a typically subtle reaction, UCC said it would review its other Belgian storage activities and might consider transferring its 130,000 ton/year storage facilities in Antwerp to Rotterdam, rather than risk a recurrence of the Malle affair, which it attributed to 'post-Bhopal syndrome'. This would cost Belgium around 150 jobs.

Early March 1985: Four people hospitalized and dozens of others reported burning in nasal passages, nausea and other symptoms when 5,700 lb of an acetone/mesityl oxide mixture were released at Institute. The director of the State's Air Pollution Control Commission said UCC waited three and a half hours before reporting the spill.

15 March 1985: A leak of 100 lb of a toxic gas from a UCC plant in South Charleston, West Virginia, spread over a shopping mall a quarter of a mile away, slightly injuring eight people.

28 March 1985: Chlorine leak at Bhopal.

1 April 1985: Chlorosulphuric acid fumes leak at Bhopal.

Early July 1985: Ten million water-melons, one-third of the total California crop, were destroyed and water-melon sales were banned in California, Washington, Oregon and Alaska following contamination of water-melons by the UCC pesticide Temik. UCC accused growers of 'flagrant misapplication of the product', as Temik's label does not list water-melon as one of the crops it may be applied to. Farmers maintained UCC was at fault, as they claimed Temik stays in the soil much longer than the 100 days UCC claims it degrades in. Farmers said they used Temik four years ago while growing cotton in four fields in Kern County, California, that were found to be the source of the water-melons contaminated with Temik. California officials did not blame UCC for the pesticide poisonings, but said they resulted from farmers deliberately misapplying Temik to save money. A total of 149 cases of aldicarb sickness were confirmed in California, with 200 or more suspected cases, and twenty-four cases outside California were confirmed.

August 1985: Leak of aldicarb oxime at Institute.

May 1986: Scientists at the US Centers for Disease Control expressed fears that methods for testing foods for low levels of pesticides such as aldicarb may not be developed enough 'to pinpoint tiny amounts of illness-causing pesticides'.

June 1986: UCC agreed to pay $33.3 million to industrial-gas users to settle price-fixing charges.

— UCC was criticized by a West Virginia government agency for failing immediately to notify state authorities when ten tons of ethylene oxide spilled into the Kanawha River on 25 May. The leak began at 8 a.m. but was not reported until after 2 p.m. The local state inspectors did not respond to the report of the leak for more than eighteen hours, as the local agency responsible does not have money to pay for weekend inspections.

September 1986: The Occupational Safety and Health Administration cited UCC for twenty alleged violations at their plants in South Charleston and West Pkifflin, Philadelphia. Proposed penalties were set at $105,600.

February 1987: The Conservative Party chairperson of the British Commons Agriculture Select Committee criticized the British government for clearing use of the pesticide Temik, when animal tests had been reported in December 1986 to have shown that it damaged the immune system.

February 1987: The US Federal Trade Commission has charged UCC with violating a 1977 consent order regarding UCC sales of industrial gases.

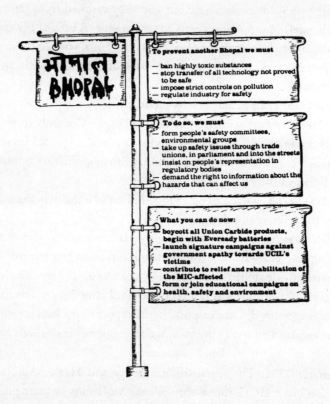

8 Accidents Will Happen

'What chemical companies do about safety problems always lags behind what they know they can do.' – Ian Swift (quoted in **Wall Street Journal Europe,** *13 November 1985*)

The Design Institute on Emergency Relief Systems has received little public exposure, even though its report revealed design faults that present toxic capital with a major new possible cost. The chemical industry has no inherent drive to improve plant safety.[1] Whether the Design Institute on Emergency Relief Systems' proposals are implemented depends on whether this is made into an issue on which toxic capital is attacked, thus pressurizing the state to regulate the industry's design standards.

Yet toxic capital's safety crisis cannot be resolved even by this enormously expensive technical fix. Within the industry there are a variety of possible failures, only some of which could be mitigated by improved emergency-relief systems (Kletz, 1985). As *Midnight Notes* reminds us, 'though any *particular* accident, by definition, can be avoided, accidents in *general* are unavoidable' (1980, p. 23). Accidents are an inherent part of the operation of chemical-process systems, as they are an inherent part of any complex system.

Indeed, accidents in chemical-process systems are guaranteed by the design of these systems. The design of chemical-process plants is determined not only by technical and economic factors, but also by management strategies that have grown up inside the industry. Contradictions regarding the availability of information to plant workers are one crucial factor that determines the course of accidents. Other factors operate as well. Among them are the increased technological complexity of the plants, with a concomitant increase in the possibilities of failure in different parts of these complex systems, and the continued expansion of the industry, which produces more plants at which these failures may occur and thus increases the likelihood of such accidents.

The industry response to the safety questions raised by Bhopal may be summarized as follows:

The industry is one of the safest in the USA, safer even than banking or housework. Although the industry poses risks, it is concerned with them and is constantly increasing safety by improving safety technology; that is, safety in the industry is technology-driven. Bhopal was a rare occurrence and, besides, that was in a Third World country. Higher safety standards are assured in metropolitan countries by government regulations, better workers and a culture of safety. Nevertheless, some risk still remains. We have the safest systems, but we cannot yet eliminate human error. The human error always means the plant operator, the worker – never the management or the design and other engineers. The technology is as near perfect as can be. We are constantly updating it; we constantly plan for and introduce technology to deal with accidents, even though these accidents are rare and unusual.

This chapter takes issue with that industry view of risk and 'human error'.

Good Safety Record?

It is necessary, first of all, to look at the industry's safety record, though secrecy makes this difficult. For an industry so proud of its safety record, the chemical industry neither invites nor welcomes inspection. Any attempt to look at the industry's safety comes up against the blank wall of secrecy. Documentation on chemical accidents is not plentiful, given the 'reluctance of the industry to review accidents in a public, accessible form . . . The most important material is produced by the industry itself, and unfortunately, it is circulated privately and is not available to researchers outside the industry' (Perrow, 1984, pp. 102–3). Perrow later comments on a survey of vapour clouds by Davenport:

The author of the article is associated with the insurance industry, and thus has access to all the published data on such events. Yet 42 percent of the accidents noted, including some large and catastrophic ones, are referenced only as 'private communication'. No doubt these were covered at the time in local newspapers to some extent, but it is unfortunate that full technical discussions appear to be available only on a private basis. (Perrow, 1984, p. 113)

Even those researchers associated with the industry express dissatisfaction over the non-availability of information. Wilson (1980, p. 194) complains: 'A detailed analysis of accident records is hampered by incomplete reporting. This is particularly true of the "near miss" situations, where a serious accident is averted by prompt action or good luck.'

Such information as is available suggests that the industry's assurances of safety are nothing but assurances. The industry's basis for claiming to be the safest in the USA is based on an extremely narrow definition of safety.

In terms of 'lost-time' accidents, i.e. limbs lost and bones crushed, the chemical industry may well be one of the world's safest, largely because there are relatively few workers in this capital-intensive industry, and little of the kind of limb-crushing, life-devouring machinery used, for example, in coal mines or foundries. But in terms of the danger of long-term effects and catastrophic accidents, the picture is much different. (Everest, 1985, pp. 36–7)

A comment made by a New Jersey chemical-plant worker is particularly appropriate here: 'There must be something wrong. This plant has been going since 1939 and there's only seven guys of pensionable age' (Halle, 1984, p. 114).

The surest source of information is the insurance industry. A survey by Marsh and McLennan of the hundred largest monetary losses in the petrochemical industry in the last thirty years appears to contradict the industry's assurances that safety is constantly improving. This survey shows what Trevor Kletz, safety consultant describes (*Chemical Engineer*,

April 1986, p. 55) as a disturbing time distribution. The distribution of monetary losses over the past thirty years was as follows:

> 1955–64, 13 percent
> 1965–74, 28 percent
> 1975–84, 59 percent.[2]

While undoubtedly inflation contributes to this time distribution, so also does the increased size of chemical plants and increased competition in the industry. The secretary-general of the European Environment Bureau, Ernst Klatte, has said the chemical industry in West Germany has experienced more than one hundred major accidents in the past decade. Klatte argues that the chances of major accidents are rising as increased competition forces European firms to cut back on maintenance in plants built during the petrochemical investment boom of the 1960s and early 1970s (*Europe Chemical News*, 17 February 1986, p. 21).

Other publicly available records refute the industry's claimed excellent safety record. A review of reported chemical accidents in New York State from January 1983 to November 1985 found almost five chemical accidents occurred each week. The study further found the number of accidents increased 53 percent from 1983 (an average of 3.4 per week) to 1985 (an average of 5.4 per week). The 706 reported accidents resulted in nineteen deaths and 229 injuries. An EPA survey found some 6,928 major chemical accidents had happened in the USA over the previous five years. Finally, a world-wide survey by Henri Smets of the Paris-based Organization for Economic Co-operation and Development, restricted to toxic-gas leaks in the six months from September 1984 to February 1985, found twelve serious accidents, excluding Bhopal (Morehouse and Subramaniam, 1985, pp. 90–2). Not bad for just one type of accident in the safest industry in the world!

Speaking at a conference in Newark, New Jersey, in March 1985, the organizational specialist Charles Perrow estimated the risk presented by the chemical industry: 'At the present level of plants with these characteristics and volumes of production, Perrow insisted that we can expect catastrophic accidents, which he defines as those that kill or seriously injure one hundred or more bystanders, at the rate of about four a decade' (Morehouse and Subramaniam, p. 89). Perrow would generally be condemned by industry sources as a doom-monger.[3] Yet, looking at the record of 1984, his predictions appear tame. The Bhopal killing left between 5,000 and 10,000 dead, and 30,000 to 40,000 seriously injured. The liquid-petroleum gas leak at Ixhuatapec, Mexico, left 252 dead, 1,000 missing and hundreds injured. Fire from a leaking petroleum line in Cubatão, Brazil, in February 1984 left at least 500 dead. Smets's six-months analysis of major gas leaks includes four with the number of injured over 100; whether or not they were seriously injured depends on Perrow's definition of serious injury. In October 1984, in Linden, New Jersey, a pesticide leak put 160 people into hospital. In December 1984, an ammonia leak at Metamoras, Mexico, led to 200 receiving hospital treatment, and an ammonia leak in Cubatão, Brazil, led to the hospitalization of 300. In February 1985 an ammonia leak in North Sumatra, Indonesia, injured 130 workers, some of whom required hospital treatment.

Thus, even on the limited material available, industry assurances of safety are unconvincing. Some industry sources agree. In August 1986, David Brandweir, a consultant engineer, told an American Institute of Chemical Engineers meeting in Boston that errors such as leaking wastes and raw materials, underground storage tanks and uninspected containers for holding hazardous materials are 'common to all plants, regardless of size, geographical location and/or type of industry' (*Chemical Marketing Reporter*, 1 September 1985, p. 5). Brandweir's conclusions were based on surveys of eighty-eight facilities, including chemical and petroleum plants, general manufacturing and metal-finishing plants, as well as waste dumps.

A generally pessimistic view of the chances of major accidents also comes from another unlikely source. Gail P. Nostrom II, of Industrial Risk Insurers, in his study of fire and explosion losses in the chemical-process industries, notes:

Clearly, major losses in the chemical and allied classes can be characterized as having relatively low frequency, but enormous dollar-loss potential is associated with them. This could be considered evidence in support of Murphy's law: ever-decreasing frequency produces an ever-greater loss. In other words, the incident which is not supposed to happen will, with a greater loss potential associated with it than expected. (Nostrom, 1982, p. 81)

Before looking at the causal factors for major industrial accidents, it's necessary to confront the industry's continued attribution of accidents to human error. We will then consider other types of human error not normally emphasized by industry spokespeople – management error and design error. We will proceed from there to show how human error is produced by the systems within which the humans work, by management control and determination of the work process, and how management attempts to reduce human error lead to increased dangers. We will finish the chapter with an examination of the 'normal accident' or 'system accident'.

A Spectre is Haunting Toxic Capital – The Spectre of Human Error

Industry responded to Bhopal by laying the emphasis on human error. In a *New York Times* Sunday feature article on the industry's safety, the wild card is not the technology but the worker:

It is the potential for human error that continues to haunt [chemical] industry executives. Mr Hennessy of Allied emphasizes that even the 100 safety audits his company performs on its plants each year does not remove the spectre of a catastrophic human mistake. 'That's the one I live in fear of the most,' he said. 'It is the one all the chemical industry executives are living with today.'
(New York Times, *16 December 1984, section 3, p. 30*)

Similarly, Dow Chemical's John M. Jones, head of the US Chemical Manufacturers' Association emergency planning efforts, responded to the chemical accident at Kerr McGee's Gore, Oklahoma, plant in January 1986 by emphasizing human error: 'In many of these accidents, the human element is the biggest problem. A lot of people say you have to add equipment. But you can't make things idiot-proof. You still have to properly train people, have proper procedures – and stick to them' (*New York Times*, 7 January 1986, p. B5).

Blaming the workers crosses supposed ideological barriers. 'Six human errors led to the Chernobyl nuclear disaster, according to Valery Legasov of the Kurchatov Atomic Energy Institute' (Lamb, 1986, p. 43). This bosses' explanation is also used uncritically by some environmentalist groups such as Greenpeace. Responding to another accident at the nuclear reprocessing plant at Sellafield in November 1986,

> Mr George Pritchard, of Greenpeace, said that the incident again proved that no matter what precautions were taken by the company, human error would always be the Achilles heel of the nuclear industry. He continued: 'The Achilles heel caused a discharge from Sellafield in 1983, which closed the beaches of West Cumbria, it caused the accident at Three Mile Island in 1979, it also caused the Windscale reactor fire in 1957, and was responsible for Chernobyl.' (Guardian, 29 November 1986, p. 2)

This 'blaming the operator' is presented as an explanation, not only for these spectacular accidents, but also for most common accidents. The UK Atomic Energy Authority's Watson has written: 'It is well known in the case of aviation that 70 percent of accidents are due to crew error and similar figures apply to the shipping and chemical industries' (Watson, p. 323). In typical fashion, this blanket statement is neither referenced nor bolstered by actual data. Similarly Britain's Central Electricity Generating Board's L. J. C. Williams, formerly of the UK Atomic Energy Authority, and J. Wiley from the University of Technology at Loughborough attribute between 60 and 80 percent of all major accidents to human error (1985, p. 354). Perhaps the most outrageous example of passing off deeper problems as a deficiency on the operator's part, in this case a lack of discipline, is provided by Dow Chemical's D. A. Rausch, who commented on Flixborough, Seveso and Bhopal: 'In each of these cases, it was not the technology that was lacking, rather the discipline required to follow procedures and good operating practices' (Rausch, 1986, p. 14).

The value of these 'explanations' for management is obvious. The chemical industry responded to Bhopal by blaming human error, stressing the need for idiot-proof plants, yet at the same time admitting the impossibility of designing such plants. This identifies workers as the major cause of accidents. Yet it is the work process itself – the plant design, of course, but also the social patterns of command and control built into the technology – that are the major causal factors in human error. The industry's strategy of focusing on human error moves the responsibility and need for action from capital to labour.

This strategy leads to either of two responses. Either the workers must

be better trained, using simulation techniques similar to those used in the nuclear industry and being provided with more information, for example; or they must be moved as far as possible from a central position in the work process by automation and computerization. The latter allows management to argue for even greater control of the work process and increased subordination of labour.

Before looking at the implications of this ideological position, we need to undermine its factual basis. Human error, carelessness, negligence, the 'accident-prone' worker: these are capital's constant refrain when faced with accidents at work, and they have a long history (Gersuny; Nichols and Armstrong; Sass and Crook). The clichéd image of human error propagated by management runs thus: we have these wonderful systems, perfect in every part, which the workers constantly fuck up. Human error is always presented as worker error: errors by management or scientific and technical personnel are merely 'technical errors', unavoidable oversights or mistakes which are inevitable in the development of any technology. Human error includes a sub-text of blame: the workers are apathetic, careless, bored, feckless, stoned or stupid.

When we look at the reality of human error, we discover a much more complex picture. It becomes obvious that the industry presentation of human error is a tactic on its part to draw public attention away from technical and organizational problems which are more basic than operator error and which themselves produce operator error. As Embrey (1981, p. 182) notes: 'The use of terms such as "negligence" is frequently a substitute for real analysis of the causal factors which have given rise to error in particular situations.' To which we may add Lagadec's observation that ' "human error" can only constitute an excuse which permits [one] to close the case for good' (Lagadec, 1982, p. 340).

That human error is not the major cause of accidents in the chemical industry can be seen from the few data that are publicly available. Thus one Dutch study found the following to be the main causes of chemical-plant accidents: defects in design, manufacture or erection of plant equipment (15.5 percent), operational failures (49 percent), human failure (32 percent) and external events/natural causes (3.5 percent). A Japanese study (Hayashi, 1985) found human error accounted for only 25 percent of accidents from 1970 to 1979 at chemical complexes in six districts of Japan. The simplicity of the human-error explanation is also confounded by the following data on accidents in the chemical and petroleum industries from Nostrom, 1982):

Contributory Factors

Explosions causing Property Damage greater than $100,000, 1978–80	%	Fires causing Property Damage greater than $100,000, 1978–80	%
Equipment rupture	26.7	Sprinkler or water spray lacking	35.6
Human element	18.3	Human element	15.6
Improper procedures	11.7	Presence of flammable liquids	11.1
Faulty design	11.7	Vessel/equipment rupture	8.9
Vapour-laden atmosphere	11.7	Excessive residue	8.9
Congestion	8.3	Production bottleneck	6.7
Flammable liquids	6.7	Faulty/inadequate sprinkler/water spray	6.7
Long replacement time	6.7		
Inadequate venting	6.7		
Inadequate combustion controls	5.0		
Inadequate explosion relief	5.0		

This section of actual data can be concluded by noting that, even in cases where human error is listed as the main cause of accidents, this may be due to insufficient investigation or deliberately biased reporting. Consultant engineer Ken Wood points out: 'Investigations into disasters usually concentrate on the technical reasons for their occurrence. However, there are always underlying failures in management, which allow the technical failures to happen' (Wood, 1985, p. 20).

A Different Class of Human Error: Management Error

This brings us to a different class of human error, which is rarely criticized: management error. Howard (1983) provides a very useful corrective to the limited and ideologically slanted attribution of human error to workers in an analysis of five illustrative accidents, only one of which could be attributed to the classical notion of human error. It's worth quoting at length his detailed treatment of one such accident:

In one accident a hold tank ruptured with great force from an exothermic decomposition of the contents. No appreciable amount of reaction was supposed to take place in the hold tank, yet the explosion occurred at a pressure of about 20 bar. Production from the operating unit was interrupted for an appreciable time, and there was major mechanical damage and business interruption loss. Fortunately no one was injured.

For several hours there had been a gradual increase of temperature of the contents. However, the operator on duty was not aware of the full significance of the continuing temperature increase. The reason was that the operating management had put the operator in charge of the portion of the production unit containing the hold tank even though the operator had had no training for the operation. This may seem almost unbelievable, but such things do happen. In the particular unit involved there had been very poor training for a number of years. Furthermore, there had been no determination of the decomposition properties of the particular material in the hold tank. Management had some years previously felt that tests should be made because it believed that the material could exothermically decompose. However, this same management kept postponing initiation of the tests because it was 'too busy' running the unit, making high production records.

This accident was not caused by the 'failure rate' of any equipment. Nor was it due to 'human error' rate as usually thought of; i.e. it was not due to forgetfulness. Instead, it was due to another form of human error: conscious decisions which ignored safety implications. *There was a conscious decision to postpone testing of a process material. There was a conscious decision to save money by eliminating most of the costs of operator training. There was a conscious decision to put an untrained operator in charge. (Howard, 1983, p. A12, my emphasis)*

In three of the other four accidents he examines, Howard finds the cause to be conscious decisions on the part of management to ignore safety considerations, mainly for economic motives. Similar economic causes may be found behind the chemical disasters at Flixborough and Bhopal, as only two examples out of many. The continuous operation of chemical-processing plants is essentially due to the huge amount of capital invested in the plants and also due to the nature of the process itself. To obtain an adequate return from the vast amounts of capital tied up in petrochemical plants, the industry must run its plants constantly: the imperatives of production produce what has been described as a culture of avoiding down time. This means that plants are often not taken out of production for repairs and maintenance unless this is essential for production to continue: such repairs as are done may be rushed jobs (as at Flixborough) to enable production to continue again.[4]

Similarly, in times of increased competition in the industry, maintenance tasks receive less attention and inferior-quality material may be purchased. The recent economic crisis in the chemical industry has led to cutbacks in staffing, supervision, maintenance and safety; combined with concurrent government deregulation of the industry, such cutbacks increase the risks of accidents occurring. Two quick examples may be cited: the decline in valve quality and the increasing role of purchasing managers. Recent tests by Du Pont are reported to have shown that as many as 30 percent of new valves delivered to the company were defective: this is the result of an economic crisis in the valve industry. The switch from engineers to cost-conscious purchasing managers in sourcing equipment is lowering standards; one chemical-industry consultant went on record as saying that 'The single most hazardous part of the chemical industry is the purchasing agent. They don't have any idea how to judge quality.'

It's important to recognize that this cost-cutting attitude is not limited to the safety (and other) equipment central to the production process itself, nor to unprofitable plants. Similar practices can be found in highly profitable plants, illustrated by an Irish example. Merck, Sharp & Dohme's bulk-pharmaceuticals manufacturing plant at Ballydine, Co. Tipperary, produces an estimated $1 million a week profit for the company. Yet during the 1985 High Court case brought against the company by local farmer John Hanrahan, the company admitted it had not replaced a wrong-sized flow meter in its toxic-waste incineration system. It did not do this despite the fact that correct flow was important for the chemicals to be incinerated at the proper temperature and the fact that the meter itself would have cost a mere £1,000.

Merck's attitude seems to have been that operation of plant not essential to the production process may not require so much care. This was shown in the way the company operated the incinerator in violation of its maker's specifications and in disregard of the fact that incineration of the plant's toxic waste at insufficient temperatures led to increased toxic chemical emissions. The fault lay not with the operator, but with the management, which failed to inform the workers that correct temperature was essential for the 'safe' operation of the incinerator. Perhaps 'correct' would be a better word than safe here, since there is a whole debate regarding definitions of 'safety'.

In any case, the industry often operates with total disregard even for its own highly limited and dubious definitions of safety. Again, as in Bhopal, this lack of concern by local management grew from a lack of concern by the multinational's US management. The fact that for producing profits the correct operation of the incinerator was unimportant is obviously central here: we may assume operators are informed of the need to pay close attention to the correct temperature of the various production processes where such attention is essential for the production of a saleable product.

This inclusion of management error is a very useful expansion of the term 'human error', which normally means unconscious or subconscious acts, or their omission, by production workers. The term can now be expanded to include management decisions regarding plant operations which ignore safety considerations in the search for increased production, productivity and profitability. The term may be expanded even more, however, when we consider design error.

Design Error

Here again, if we look behind the easy explanations of worker error, a different picture emerges. To understand the next quotation, it's important to realize that there are two radically different types of chemical plant: batch plants, which produce batches of various different chemicals, depending on the market and production requirements of the owners, and continuous-process plants, which are dedicated to the production of one chemical only.

Batch plants are usually thought to be more frequently involved in accidents than continuous-process plants. This is normally attributed to the fact that operator decision and corrections are needed all the time.

The high accident rate for batch operations because of design error, as indicated by the study, points to the design as a major contributing factor. It appears that it is difficult to make a design which can cope with the unstable process conditions of a batch unit. (Haastrup, 1983, p. J20)

The reason that little attention has been paid to design errors is not hard to find: 'Design error in the chemical industry, or in any industry, is a problematic area to study, because the industry often seems to find it difficult to accept that the problem exists at all' (Haastrup, p. J15). As usual, information is unavailable: 'The possibility to study the problem is perhaps greater in the nuclear industry, where legislation requires rather strict records of all incidents. Such records are not generally available for the process industry' (Haastrup, p. J25).

Haastrup presents the following table on design-error accidents as a percentage of all accidents:[5]

Author	Industry	Design-error Accidents as a Percentage of all Accidents %
Taylor	Nuclear	36
Faircett	Nuclear	26
Nostrom	Chemical (fire/explosions only)	12
Ordin	Hydrogen	22
Haastrup	Chemical	25

Haastrup attributes the higher figures for the nuclear industry to better record-keeping and the lower figure from Nostrom to a different definition of design errors. Haastrup's survey of 215 cases found 317 design errors, an average of 1.5 errors per case studied (p. J22). Of the cases studied:

It is interesting to note that totally unknown phenomena are recorded in only three cases (1 percent). This is contrary to what is found in the nuclear industry, where figures like 25 percent were found. A possible explanation is that the cases from the chemical industry do not offer sufficient information to judge this question. Another possible explanation is that the nuclear industry is a newer technology than the chemical industry. (Haastrup, p. J24)

I would incline towards the lack-of-information explanation, especially given the industry tendency, noted previously, to present only enough information to provide a reasonable and plausible explanation for an accident, rather than a correct one. Nevertheless, it must be noted that Dow Chemical's D. A. Rausch reported, 'Over the past five years, greater than 98 percent of our reactive chemicals incidents have been related to known chemistry. It is not the unknown that gets us, it is what is known or what has already happened to others' (Rausch, 1986, p. 15).

The idiocy of some design errors is hard to credit. Two examples from many should suffice:

Software had been written to display information 'geographically'. Three different visual-display units contained information from three different areas of plant; each visual-display unit 'page' was organized around a specific location of the plant ... this decision required the operator to examine information from a variety of different locations. To make a decision the operator had to move backwards and forwards between the three visual-display units and within each visual-display unit he had to locate different 'pages', then search each 'page' for the information he required. This is a very tedious procedure and one which is likely to cause error as the operator tries to remember information so far collected, whilst concentrating on securing new information. Thus a design modification developed for sound engineering reasons creates much more difficulty for the operator than the conventional displays it is replacing. (Shepherd, 1982, p. 153)

Again it appears that it is difficult to make a design which can take care of changing conditions in the process plant. The high accident rate for maintenance and repair (20 percent) found in this study indicates the need for a design checklist with special emphasis on these problems. A typical example is chemical burns to maintenance personnel when they open a line which cannot be drained because a drain is missing. (Haastrup, p. J21)

This examination of design error is not only intended to expand the concept of human error into the technical apparatus. These design errors are inevitable given the economic pressures at the design stage and the limited knowledge designers have of many chemical processes. Thus there are both economic and technical causes for these design errors. On the latter question, Haastrup (p. J23) is again extremely enlightening: 'In approximately 35 percent of these cases it was judged that the designer had an insufficient model of the process conditions under which the plant was to be run. A typical example is that corrosion was not taken into account.'

The reasons for design errors in the chemical industry are not hard to find. They vary from the use of untried technology, through scale-up problems, through actual ignorance of what takes places in the chemical process, through to economic pressures due to the position of the industry or firm. The importance the industry has previously placed on safety training for the education of chemical engineers can be seen in the fact that none of the 145 undergraduate chemical-engineering courses in US colleges requires separate safety courses as part of its curriculum. A survey of engineers, including chemical engineers at a major US chemical company, found 'nearly three-fourths of respondents had never taken a college course in which safety was a major topic. Although 11 percent had taken two or more safety-related courses in college, most doubted their utility. Only slightly more than a third felt their college education adequately prepared them to deal effectively with safety issues they routinely faced on the job' (Flores, 1983). A more recent study for the US National Institute of Occupational Safety and Health found: 'Preliminary results indicate that many accredited engineering programmes make no claim to cover safety and health issues for their students' (Talty, 1986, p. 15).

An indication of the poverty of engineering-safety education can be found in Talty's recommendations, published in October 1986 – ten years after Seveso, two years after Bhopal, a year after Institute: 'There is a need to strengthen existing accreditation criteria to clarify the fact that safety and health in the work-place is an engineering issue that requires the attention of the engineering community, both in the classroom and on the job . . . The development of a full understanding of, and commitment to, the maintenance of safety and health *must become* an integral part of the education of every engineer' (Talty, p. 16, emphasis added).

Given the lack of such education, it is no surprise that safety is not a major determinant of chemical-plant design. While safety may not figure largely in engineering courses and textbooks, economic analysis does. One text, which begins by stating baldly: 'The role of the process engineer varies according to his employment. His objectives are the same as those of his company and are primarily economic . . . the basic strategy of the process engineer should permit the adjustment of the technical aspects of the design to produce the most profitable solution' (Wills, 1973, pp. 3–4). This economic objective is paramount to all other considerations, including good design practice: 'Assuming the objective to be optimized is economic, if good practice decrees an excessive time on design, thereby increasing costs, or alternatively creates delays in which rivals may bring out processes or obtain sales contracts, then good practice itself will vary' (Wills, p. 111).

L. N. Davis gives a wellnigh incredible description of some of the pressures that affect the design of chemical plants:

Finally, there is the time element. 'In industrial works', note industry experts John Happel and Donald G. Jordan in their book, Chemical Process Economics, *'it often happens that estimates of new capital investment and manufactured cost must be made in a very short time – say eight hours.' While it may seem ludicrous to devote so little time to decisions involving a multi-million-dollar enterprise, it occurs so frequently that engineers have had to learn to live with it. They must be able, for one thing, to make an accurate design of a distillation column or a chemical reactor in a quarter of an hour. 'This lack of time in which to complete the work is often accompanied by a lack of knowledge concerning the physics and chemistry of the process,' the authors go on. 'This lack must be made up by a combination of experience, knowledge and judgement.' It is not humanly possible to understand everything about a given process before the plant commences operation. (1984, pp. 187–8)*

Given such economics-derived pressures on the basic plant-design process, it is no surprise that safety holds such a minor place in the basic design of chemical plants. Thus, when safety and other specialists are consulted, they have to deal with a design which has already had much capital invested in it.

Hazard and operability studies come too late for fundamental changes in design. All we can do is add on safety equipment to control the hazard, thus producing extra complication and expense. If we could provide more time in the early stages of design for consideration of alternatives, we would be able to avoid a lot of our hazards . . . However, progress towards these 'inherently safer plants' has been disappointing. (Kletz, 1985b, emphasis in original)

It's necessary to stress the lack of knowledge Happel and Jordan mention. Perrow (1984, p. 85) emphasizes that the industry's understanding of what goes on in many chemical processes is limited: all the engineers may know is that the process works, not how it works.[6] This, almost traditional, ignorance is joined by scale-up problems due to the increased size of plants, caused by economic factors. Rasmussen explains:

Another aspect of modern industrial activities is the short time-span from product or process development to large-scale operation, which does not permit the evolution of safe systems from feedback of operational experience. The consequence is that for safe design of industrial systems, empirical design rules and equipment standards are gradually being replaced by explicit risk criteria and analytical techniques for safety-equipment assessment. (1982, p. 143)

It is not reassuring to be reminded: 'It is interesting to note that the current approach made in the design and evaluation of these systems is very similar to that in nuclear-power plants' (Green, 1983, p. 13).

There are major problems here. MacKenzie notes: 'It is difficult to guess adequate safety measures in advance. Experimental simulation of every possible accident can be more expensive than the equipment that engineers want to protect' (1986, p. 43). The analytical techniques and extrapolations can be incorrect: 'Engineers try to predict how pressure rises during a runaway reaction by extrapolating their observations of normal reactions. They are often wrong' (MacKenzie, 1986, p. 43).

Halle presents a shop-floor example of possible effects of chemical engineers' ignorance of chemical process and safety:

The new tryamine plant exposed operators to an additional danger – untried technology. For instance, after a year of operation an urgent directive came from company headquarters. Their research chemists had realized that a combination of ingredients used in the plant was highly explosive and should be discontinued at once. Such are some of the benefits to operators of working on the frontiers of new technology! Compare the dubious privilege of being an operator in a nuclear power plant. (1984, p. 114)

One final example of technical ignorance will be cited: 'There are instances where a manager or an engineer providing the technical input to a task analysis, that is, specifying what the task involves, is not clear himself about how parts of the plant work or how they should be operated' (Shepherd, 1982, p. 154).

Producing Human Error

So far the simple image of human error as the main cause of accidents in the chemical industry has been undermined by a look at actual information on the causes of chemical accidents. A further step brings us to see that much human error is actually produced by the systems design. As Embrey (1981, p. 182) notes, 'Extensive studies have shown that most human failures are induced by the unsatisfactory design of the system within which the human being operates.'

The major example of relevance here is the effect of continued automation of process plants on workers' alienation from their work. This increased alienation makes it more difficult for workers to intervene in the work process when something goes wrong. Automation removes the operator from the system and makes the actual chemical process almost unreal to the operator.

The situation arises that there is a lack of process-related activities for the operator; he is insufficiently involved in the operations of the process. This may evoke the feeling that he can no longer adequately control the system, he may feel uncertain in case of manual control actions. The result can be a situation in which the operator no longer feels that he can contribute in a meaningful way to the process operations. As a monitor he waits for irregularities in the process. (Brouwers, 1984, p. 193)

There is more involved here than workers' *feelings* of alienation. There is actual alienation. The *Wall Street Journal Europe* (28 January 1985) reported on one worker's experience of the work process at Du Pont's Deepwater, New Jersey, phosgene plant: 'But so little happens most days when he works in the control room, Mr Hermann says, that he usually just paces back and forth for eight hours, surveying the instrument panel.' Let's supplement this authoritative source with two quotes from workers at a chemical plant in Elizabeth, New Jersey: 'Actually this is a pretty boring job. When things are going smoothly there's nothing to do. All it really is is watching dials and gauges. Well, I read and stuff.' 'In the beginning, it

CMA photo assures us that computerized control panels 'track processes, conserve energy and minimize waste': enhancing safety or producing human error?

was interesting, but after it got dull. It's the same shit, day in, day out. It's boring. There's no variety' (Halle, p. 107). The sociologist who studied these workers, David Halle, was himself quoted by the *Wall Street Journal Europe* as comparing life in an automated plant with 'being a soldier on guard duty in an area the enemy rarely enters'.

The increasing automation of chemical plants, as a response to economic crisis and as an attempt to increase management control of the work process, will continue to reduce workers to passive monitors. 'Next to that, it is conceivable, although in the process industries not at short notice, that *all* process situations in a system that are thinkable can be controlled at the supervisory level' (Brouwers, 1984, p. 184). With this, the operator will be moved farther and farther away from the work process.

It can be deduced from the aforementioned that the near future allocation of tasks between man and machine may become even less favourable from the point of view of human needs and capabilities (e.g. an increasing lack of balance in the work-load). The number of situations in which interventions in the process operations by the operator are needed will further decrease. To the operator the complexity of such situations, however, may increase (control skill, man-machine communication). More and more the situation will be that the operator – as a monitor – has to wait for irregularities in the process operations. (Brouwers, p. 187)

This removal of the operator from direct contact with and control over the process carries two implications. In the first place it means that accidents will not be caused by operators, since they have been removed from control over the process operations. Secondly, it shifts the possible area for human error: 'The risk taken with human operators will not be that they will cause accidents but rather that they may not succeed in preventing them' (Rasmussen, 1982, p. 147).

The major reason they may fail to prevent accidents occurring is their alienation from the work process. The chemical engineering and ergonomic literature points out that it is difficult for operators to maintain a mental model of the process given their distance or alienation from the process system. This results in a lack of 'hands-on' experience and knowledge of the plant itself, losing the operator's skills. Unfortunately for management the skills thus lost are the very skills required to deal with an emergency.

Mike Cooley puts the situation this way: 'If operators are to be reduced to unthinking machine accessories lacking an overall view of the operating system with which they are dealing, then they are ill-equipped to deal with an emergency' (*Design*, October 1985, p. 23). This is just one more example of a major contradiction at the heart of the labour process: capital needs workers to act as unthinking machines and yet also depends on their creativity.[7]

As more inner loop functions get automated, the operator is required to function more at the outer loop stages. A problem occurs when an inner loop function fails and the operator is required to jump in the loop and find the failure ... Kessel and Wickens and Young found that operators who developed an internal model of the system from monitoring the system were not very good at detecting system faults. Operators who developed an internal model from controlling the system were much better at detecting system faults. (Eberts, 1985, p. 34)

Woods's examination of operator behaviour in emergency events noted: 'In general, when operational problems did occur, it was because operator state identification and control activities gradually became decoupled from actual process state as a function of execution problems or the unexpected' (Woods, 1984, p. 22). There are technical problems with realizing the difference between how operators think the process is going and the actual state of the process: 'When there is some mismatch between actual and perceived process state it is difficult for those involved to modify their view given today's control rooms.'

The classic example here is Three Mile Island.

Consider the situation. One hundred and ten alarms were sounding; key indicators were inaccessible; repair-order tags covered the warning lights of nearby controls; the data print-out on the computer was running behind (eventually by an hour and a half); key indicators malfunctioned; the room was filling with experts; and several pieces of equipment were out of service, or suddenly inoperative. (Perrow, 1982, p. 180)

Consider the situation in Bhopal. At the time of the leak there was only one worker in the control room. He found it virtually impossible to check all the seventy-odd panels, indicators and controllers. Then, again, Agnew reports, 'The pressure indicator control was also known to be faulty and was therefore always taken to be suspect with its readings . . . the temperature indicator alarm had been giving faulty readings for years and none of the operators took its figures as gospel.' Simple bureaucratic work organization meant operators did not record the temperature of the tank that leaked, and thus were unaware of the significance of the rise in temperature that accompanied the runaway reaction. 'For a very long time we have not watched the temperature. There was no column to record it in the log-book.' (See ICFTU, 1985, p. 9; Morehouse and Subramaniam, 1985, p. 15.)

The problem is made more intractable by the chemical industry's obsession with secrecy. Workers receive very little information on the actual chemical process. Following a malathion leak at an American Cyanimid plant in New Jersey, a study conducted by the company found that its training methods gave workers little information on how chemical reactions occur. Therefore workers cannot always detect a contaminated process or realize how a process is going wrong or understand the implications of various mistakes.

While actual operator interventions in the chemical process decrease, the complexity of required interventions increases. The only major way suggested in the literature to deal with this problem is increased training, using simulation. For instance, Eberts's solution to operators' lack of hands-on experience is to give the operator control experience, but only through the mediation of a simulator: 'Having the operator control the process in a simulator may be an important way to develop and maintain cognitive skills that may be lost due to automation' (Eberts, 1985, p. 34). This position not only begs the question, but can actually heighten this contradiction in the work process of the process industries. 'This may add to the discrepancy between what people expect of their job and what everyday practice can offer' (Brouwers, 1984, p. 187).

Again, in many cases, training is simply an inadequate method of dealing with the problem. Some training analysis finds that the problems involved require, not training, but plant redesign or design modification. 'However, at the stage at which task analysis is carried out in order to develop operator training, managements are reluctant to make basic changes to plant and instrumentation, on the grounds of excessive cost' (Shepherd, 1982, p. 151).

The problem the operator faces is sometimes reduced to the problem of maintaining vigilance.[8] A. Craig, of the British Medical Research Council Perceptual and Cognitive Performance Unit, notes:

Another explanation suggests that vigilance declines because the work-place does not provide a sufficient variety of stimulation to maintain the brain's arousal at a level that is adequate for efficient functioning. Thus providing music, engaging the operator in brief occasional conversation or getting him to stand and stretch from time to time, have all been shown to reduce detection failures. (Craig, 1984, p. 37)

A more robust and realistic suggestion as to what to do with the bored worker is provided by ICI's J. Faulkner; he notes that, due to 'the overall reduction in manning levels throughout the industry', management has restructured operators' jobs: 'As a result, for example, Shift Managers look after larger areas and Senior Operators take on some management responsibilities, becoming Group Leaders whilst still carrying out their duties as operators. As a consequence they don't have time to be bored for long.' Indeed, in green-field plants where union restrictions are relaxed, operators can even handle 'surplus plant maintenance tasks' (Faulkner, 1984, p. 235).

Struggle for Control

Reading the chemical and human-engineering literature, one receives two different images. From the ergonomics and human-engineering specialists, the picture of the operators is of beings totally under control of the system, a pale reflection of the technology around them. From the actual plant managers one gets a different impression: workers are the wild card, dangerous, given to insubordination and shirking. The management view is closer to reality. The state of neutral peace and quiescent workers described in the ergonomics literature does not exist: chemical factories are no more immune from the class struggle than any other factories.

We have to see this whole question of human error and safety in the context of the daily struggle between workers and management for control over the work-place and the work process. A strong example is provided by a study of chemical workers in Elizabeth, New Jersey, by the sociologist

David Halle (quoted earlier). It may be argued that the situation at this plant is atypical, as the plant has been in operation since 1939; struggles over control of the work process may not be so highly developed in other chemical plants, particularly in green-field areas and in developing countries. Nevertheless some of the most advanced struggles over control of the work process have been reported from green-field sites.[9]

Halle's study is useful, in the first instance, in showing how even such supposedly highly automated chemical-process plants depend on human labour.

Accounts of automated plants usually imply that the work is all watching dials and gauges. But this is misleading, for even jobs that mostly involve monitoring do require manual work. Some valves need to be opened by hand outside the control room, either because they are too important in the production process to be left to an automatic switch that might not work or because the automated switch is broken. Sometimes chemicals leak on to the floor and must be cleaned up. And every so often parts of the machinery must be cleaned out from the inside by operators. Sometimes when I explained to process workers that I was interested in studying a highly automated plant I would get an ironic laugh: 'Automated plants! Ha! You're in the wrong place. This isn't automated. I had to go out just now and put in a new valve by hand.' (Halle, 1984, p. 104)

More important, however, is Halle's presentation of how important the workers' knowledge of the plant is to actual production. Workers have three types of knowledge which management does not possess. First, workers know the individual behaviour of machinery and equipment better than management. This knowledge comes from immediate experience and also includes all the many modifications to machinery and plant which accumulate over time with the result that 'the equipment moves a considerable distance from the design'.[10]

Most significant for the question of safety is the second type of knowledge workers possess. This relates to 'the grey zone between normal operations and catastrophe'. When problems arise, workers must decide whether to shut down production – thereby quite often destroying a valuable batch of product – or continue operations. Of similar importance is workers' knowledge of when the production process is entering a dangerous phase.

The workers' practical experience is of major use in both these areas, of greater use than management's manuals:

The instruction manual may state that certain levels of pressure and temperature are dangerous, but given the variability of equipment even on installation and the modifications it undergoes over time, this information is unlikely to be precise. It may be possible to operate quite safely within a formally designated danger zone, and there may be danger areas that are not formally designated. (Halle, p. 120)

The final type of knowledge relates to workers' actual intervention in the work process to their own benefit. Workers can intervene, especially in batch process operations but also in continuous processing, to speed up or slow down a particular process. Workers will, by accidental discoveries or by experiments, change the process, occasionally leaving out whole stages:

Sometimes men discover they can eliminate an entire stage without any apparent difference in the product. If the stage involves inconvenient work and the laboratory technicians do not notice its absence, then men may omit that stage in future batches. In many cases workers come across such devices by accident. A chief [supervisory worker] forgets to do something the formula card says he should do – cooling slowly or adding an ingredient – and the laboratory does not comment, so the 'accident' becomes a technical discovery, a useful component of the men's practical knowledge. An assistant comments: 'The formula tells you a certain material is necessary. Perhaps you're supposed to add twenty-five pounds of cobalt. And one day the chief forgets to put it in, and it works OK.

'If it's a pain in the ass to put in the material he won't do it next time. The fucking thing is cooking at 250 degrees and they want you to unbolt a manhole and put in some garbage – it's dangerous.' (Halle, p. 121)

The aim of the workers here is the same aim as has been involved in most struggles over the work process: the reduction of work time and the creation of free time for workers at work. This struggle leads to an immediate takeover of the chemical engineers' design as soon as it is placed in the workers' control:

The drive to increase the proportion of free time at work begins as soon as a new product is introduced. When a resin is to be produced that has not been made in the plant before, one of the chemists from Imperium's laboratory in Sterling Forest comes down to supervise the process. A chief commented:

'Some guy in an experimental laboratory in Sterling Forest figures it out over a pressure cooker. When they get it here the guy will stay for one whole cook. Then the next time you do it you're on your own. You start to look for easier ways to do it.' (Halle, pp. 121–2)

This has obvious implications for management. Halle's description presents a situation in which management has ceded a certain amount of control over the work process in return for a guaranteed product. The workers' struggle here is served by the nature of the technology, in particular in the batch plants:

A crucial factor that helps men impose their own rhythm on the work is that no two batches ever take the same time to cook, even if they are from the identical formula. The same alkyd can vary by several hours in the time needed to prepare it, because the ingredients are never of exactly the same quality. Ingredients that are nominally identical vary in strength and consistency. Two fifty-pound loads of cobalt will not have the same potency. This gives a certain unpredictability to production. As an assistant put it:

'Every batch is like a newborn baby. They are all different, with different problems.

'A particular process might usually take two hours, but if it takes four hours on a certain cook, supervision cannot be certain whether it was because workers dawdled, reducing the heat or the quantity added, or whether the material was below strength.' (Halle, pp. 122–3)

Thus workers can speed up or slow down the process. To quote one of the workers:

The chiefs know so much the supervisors don't know. You ask any supervisor. If it's Friday they'll [workers] get a batch out in four hours that usually takes ten. Sometimes supervisors try and run a batch straight down the line exactly like it says in the formula, and it'll take three days. And the guys here can do it in three hours. It's not like the formula says.

Every supervisor knows the guys on Friday run a batch that usually takes ten hours in four, but they don't dare say anything because next time the guys will take till Saturday afternoon and then the supervisor gets called into the office: 'How come this happened?' and all that, and he's sweating. (Halle, p. 123)

This description also involves a different view of what's involved in the increased automation of the process industry. While management can maintain that automation will increase safety, the primary reason for its introduction lies in management's attempt to regain control over the work process. As Paul Cardan puts it:

[Management] also proceeds by giving an increasingly pronounced class twist to technological development. Machines are invented, or selected, according to one fundamental criterion: do they assist in the struggle of management against workers, do they reduce yet further the worker's margin of autonomy, do they assist in eventually replacing him altogether? . . . Technology is predominantly class technology. No British capitalist, no Russian factory manager would ever introduce into his plant a machine which would increase the freedom of a particular worker or of a group of workers to run the job themselves, even if such a machine increased production. (Cardan, 1969, p. 9)

Similarly, the calls for 'increased discipline' among workers relate more to an attempt to obtain more labour from the workers than to a desire that the workers should put safety above all other priorities.[11]

In fact, automation is likely to increase danger. It removes much of the process previously under control by the workers from the workers, so that those who are most aware of when machinery and plant is entering a dangerous phase no longer have this experience and control. Given that the automated process will be run on models developed by chemical engineers and computer specialists rather than on actual experience of the process, such as workers possess, certain dangers will go unrecognized by the automated process-control system, while experienced workers will recognize them. Left to themselves, workers are highly unlikely to indulge in dangerous practices, as they are the first group affected by any accident. It is generally in response to management pressures – to keep the line going, to avoid down time – that workers are forced to undertake dangerous practices.[12]

Thus any introduction of new technology by management is not neutral. It is inevitable that management will use future changes, introduced to increase safety, to further their control over the work process and the labour force. Of some interest here is the recent proposal to introduce drug testing of workers. The chemical industry is going with the trend in the USA by introducing such testing. These programmes are being introduced accompanied by a rhetoric of dealing with the problem of drug-taking at work, with a major emphasis on the question of safety. According to a spokesperson for Du Pont: 'We recognize we're part of society and we've been seeing the drug issue getting more and more attention in society as a growing problem. We believe it is extremely important for our employees, for their protection and for the protection of the community, to be working at the top of their form' (*Chemical and Engineering News*, 2 June, 1986, p. 13).

The centrality of these safety concerns has been questioned by the American Civil Liberties Union's Adler, who told *Chemical and Engineering News:*

What has happened is because of the failure of law-enforcement agencies' efforts and public education to persuade large segments of the population not to use these drugs, and to interdict the flow of drugs into the country. Employers are being enlisted as an adjunct to law enforcement of certain – what some people might view as moral – standards with respect to controlled substances and yet it is all being cloaked in the guise of concern for health and safety in the work-place. (2 June 1986, p. 14)

Of course there is more involved here than this. It is also in management's interest to introduce drug tests. Estimates of the cost of illegal drug use to US industry vary from $50 to $100 billion annually because of lost productivity and increased health care. A Confederation of British Industry report on drug use at work argues that the increasing use of 'sophisticated' (read high-capital) equipment means workers can now do greater damage to the company and other workers than before: 'A woman overdosing on tranquillizers can do less harm on a typewriter than at a computer terminal. A man "stoned" in charge of an automated plant will cause vastly more damage than a man at a single lathe' (quoted in *The Times*, 22 September 1986, p. 1). The report says those particularly vulnerable to drugs include senior managers, salespeople, process operators and line managers.

Drug testing has been enthusiastically welcomed by US companies. It has been estimated that half of the Fortune 500 companies will have some form of drug testing in operation by the end of 1986. 'A survey of two dozen chemical firms of various sizes by *Chemical and Engineering News* finds that about one-third have some sort of drug-testing programme in place. Most are large companies . . . Safety is usually cited as the primary motive for instituting a drug-detection plan' (*Chemical and Engineering News*, 2 June 1986). These tests are generally being introduced to the most vulnerable section of the working class – the unemployed who are applying for jobs. Most companies now include such a test in their selection process or as part of a pre-employment physical examination. They are being introduced into the already hired work-force only slowly: at Du Pont for instance, they are used only 'for cause' – 'that is, when a worker has been involved in an unusual accident or has been behaving peculiarly'.

The introduction of drug testing is taking place under the guise of increasing safety. However, its main aim is to increase management control of workers.[13] It also allows management to dispose of undesirable,

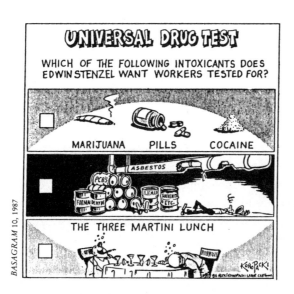

> **BE ALERT, REPORT SECURITY INFRACTIONS**
>
> **Wyandotte Corporation**
> Geismar Works
>
> # SECURITY BULLETIN
>
> To: All Employees Date: September 12, 1985
>
> From: L. E. Kelly Ref.: Drug Use on Site
>
> All Site personnel are reminded, and cautioned, to be alert for drug abuse symptoms on the Site. We continue to see evidence of illegal use during gate checks, and <u>it is quite probable that persons are working under the influence, or actually using drugs, on the Site.</u>
>
> During the past weekend, marijuana cigarette butts were found in two vehicles during routine random inspections. Four were found in one car and five in another. These were short-term contractors entering the Site in their company cars. These persons were turned over to the Sheriff's Office for handling and were also barred from the Site.
>
> These examples show that drug use can, and does, happen. It is only the careless that we catch at the gate inspections. It is up to all of us to be alert to the possibility, and to help correct, a bad problem. It is for the safe well-being of us all.
>
> Larry E. Kelly
> Security Supervisor
>
> :jc
>
> POST 9/12/85
> REMOVE 9/27/85

lackadaisical and troublesome workers. Its actual contribution to increasing safety is likely to be small and, given the likelihood of incorrect results, could be used as a method to blame workers for accidents.

Technical Fixes

Let us consider two major technical fixes to increase chemical-plant safety that have been proposed: the ergonomics fix, which calls for better plant design and includes human factors from the beginning, and the major engineering fix, which removes the human from control in so far as is possible.

The ergonomists suggest that design, particularly of control rooms, should include human factors from the beginning. Indeed, radicals suggest that operators (experienced, of course) should also be involved in the design team. This integrated approach to what are described as 'man-machine systems' would ensure the human component is considered at the same time as the machine component. This technical fix has not been embraced by the industry:

A first question is whether the integrated course is pursued frequently in actual system design. Regarding this no data are available. From the number of publications over the past decade, in which the integrated design is proposed, it cannot be supposed that such an approach has a widespread use in practice. (Personal experience in industry affirms this assumption.) (Brouwers, 1984, p. 188)

This gap between available and recommended design techniques and those actually current in the industry is confirmed by Shepherd. 'That human factors need to be considered earlier in the design of process plant is a view frequently expressed by ergonomists but still rarely heeded' (Shepherd, 1982, p. 153). According to Bellamy:

Regrettably, the impetus for improvements [of this kind] in high-risk systems has often arisen in the past only after the occurrence of major accidents. It was the investigations of the 1979 Three Mile Island nuclear accident that first drew public attention to many of the problems that can arise from poor man-machine interface design (displays and controls) in control rooms. Scientists and psychologists who specialize in studying human factors had, however, been aware of these problems for quite some time. There clearly existed a gap between 'human factors' expertise and its application to process control. (Bellamy, L., 1985, p. 27)

The reason for this gap is not hard to find: 'Unfortunately, human factors have not been particularly successful in demonstrating the cost effectiveness of the discipline' (Human Reliability Assessment Group, 1985). Indeed, consideration of human factors at a later stage in the design adds to possible costs. When problems with human factors are discovered, 'managements are reluctant to make basic changes to plant and instrumentation on the grounds of excessive cost' (Shepherd, p. 153). Consideration of human factors represents a possible cost to capital without a definite measurable benefit.

The major technical fix suggested is the engineering one: increased automation, the removal of the human operator as far as possible from control of the system:

Because the human operator can contribute significantly to the failure of complex systems, the engineer's ideal would be to design the human out of the system altogether by introducing more automatic control. This may actually make the system worse by increasing its complexity still further. In large, complex, highly automated systems the human operator becomes less of a controller and more of a monitor and diagnostician of a system whose complexity can render the effects of unexpected multiple failures difficult to predict. (Human Reliability Assessment Group, 1985)

This attempt to remove human error by increased automation is bound to fail. The most obvious reason for this is given by Embrey (1981, p. 182): 'The effect of increased computer control is often to move the locus of human error from the operational phase to the software design phase. The failure of one of the earlier Apollo space missions was due to a point being placed in the wrong position in a Fortran programme statement.'

Even ignoring this problem, there are major problems inherent in the continued automation of chemical plants, not least in the development and use of expert systems.[14] The introduction of expert systems is currently being recommended, for example, in response to the Chernobyl disaster (Lamb, 1986). Similarly, Dr Janet Efstathiou, a lecturer at Queen Mary College, London University, has predicted that intelligent machines could be assembled in England within the next decade to help prevent industrial accidents. Among the advantages of such machines are that 'they would never get bored or complacent about their performance' (*Financial Times*, 4 September 1986, p. 8).

Despite the propaganda of its proponents, however, this technical fix is still at a toddler stage:

The success of expert systems is difficult to determine. Although an increasing number of expert systems have been designed, few have been incorporated in the field. Generally, finding a metric for success has been difficult because situations can always be found where the expert system will not work. Expert systems are usually demonstrated in a limited domain and determining the extent of application is difficult. (Eberts, 1985, p. 31)

On the present development of these systems, Eberts (p. 31) notes that MYCIN, one of the most successful expert systems, contains 450 rules. In comparison, master chess players were found to recognize 31,000 basic or primitive positions, while expert computer programmers were estimated to have between 10,000 and 100,000 rules for use in programming.

The underdevelopment of this technical fix is confirmed by the inconclusive conclusion to a study by W. R. Nelson, of the US nuclear company E, G & G (Idaho) Inc., and J. O. Jenkins, of the US Nuclear Regulatory Commission: 'Much research remains to be accomplished before the true value of expert systems for the control room can be clearly established' (Nelson and Jenkins, 1985, p. 29). This study, intended to compare operators' responses with the aid of either an operator-controlled or autonomous expert system or without such a system, concluded: 'The unaided human operator performed as well as the human with either type of response-tree aid.' One possible reason for this was 'the expert system did not appear to provide assistance fast enough' (Nelson and Jenkins, p. 28)!

The more general question, which will be returned to later, is whether laying another layer of fault-ridden technology on top of an already faulty technology can in any way be considered a solution: this path seems likely only to increase the number and variety of possible failures.

One further example of an inappropriate technical fix should suffice to round off this section. One major problem relating to hierarchical systems of organization in the work-place is the effect of the hierarchy on the transmission of information. Information going up a hierarchy is more likely to be diverted, delayed or not sent at all – as anyone with personal experience of a bureaucracy can confirm.

In nine out of ten major accidents studied by L. J. Bellamy, nine involved interpersonal communication errors, while 'in two of the accidents analysed the frequency of identified communication errors exceeded ten' (Bellamy, 1984). Hilariously, Bellamy's solution is to cut out the human as 'imperfect' and to propose adding another layer of complication to the system: 'Finally, it should be apparent that natural languages are a highly imperfect means of communication as far as systems safety is concerned. Future research should perhaps concentrate more on designing artificial language systems which are not prone to the problems of the alternative open systems currently in use' (Bellamy, L. J., 1984, p. 175).

It is only fair to mention here a reformist trend towards increasing chemical plant safety among chemical engineers. This trend is best exemplified by the work of Trevor Kletz, ex-ICI safety manager and a safety consultant to the industry. Kletz wishes to decrease chemical-process hazards by designing hazards out of the plants and processes at an early stage. His response to a chemical as hazardous as MIC is: don't use it; substitute a different chemical and, if necessary, a different process. This position clashes headlong, for example, with the economic interest of UCC, whose proprietorial interest in MIC makes it unlikely ever to abandon the chemical, unless forced to do so.[15]

When use of a highly hazardous chemical is unavoidable, Kletz's response is that less of it should be used. Again in the case of MIC, this clashes with UCC's role as supplier of MIC to other pesticide manufacturers: this role means UCC must store MIC. Kletz's campaign for inherently safer plants has not been a rip-roaring success, given the industry's refusal to put so high a value on safety that it becomes basic to plant design. Following Bhopal, Kletz's prescription to use fewer hazardous materials has been adopted by many chemical companies.

Accidents are Normal

The increased complexity of chemical-plant design increases risks. The speed at which new technology of increasing scale is introduced in chemical processing mirrors the speed at which new toxic chemicals are introduced without adequate testing. Assurances are given that safety is assured by risk and reliability analyses and by the increasing use of redundant safety features – assurances that echo those of the nuclear industry. Tools such as risk analysis and reliability analysis generate quantifications to reassure us. These are again reminiscent of the nuclear industry's monologue on safety. Let's note a few practical problems before we get to the heart of the matter.

To begin with, risk analysis does not always work: its application and usefulness are limited. To quote a chemical engineer:

These quantitative risk analyses have proved to have value in a limited number of cases but only in an approximate manner. It should be noted that, after one very special risk analysis, and after all safety measures indicated by the analysis had been installed, a significant accident, unpredicted by the analysis, occurred. This is by no means a criticism of what was in fact a very good risk analysis, but does indicate limitations of risk analysis. (Howard, 1983, p. A14)

This chemical engineer accuses those who are enthusiastic about the use of risk analysis in safety of becoming engrossed in academic discussions, mesmerized by technique for the sake of technique. Howard compares this with sophisticated medieval theological arguments over how many angels could dance on the head of a pin. He concludes 'the expertise, finesse and sophistication of risk quanitification procedures outstrip their accuracy and utility' (Howard, p. A17).

Reliability assessment also faces problems. The major one relates to lack of information on human reliability. (This appears strange, given the industry's constant reiteration of the importance of human error in accidents.) There is a lack of basic data for these probabilistic risk and reliability assessments. Reading the literature, one comes across the same complaint continually. 'There is an increasing cry in most countries, from designers and assessors of highly reliable systems and other associated parties, that the appropriate data are not available' (Green, 1983, p. 94). 'The dearth of "real" data is a stumbling block which needs to be overcome' (Watson, 1985, p. 335). Although a quarter of human errors are supposed to involve maintenance failures, 'With the exception of the analyses of Joos *et al.* (1979) which have been converted to a postulated error rate based on the likely number of operations per year for a given plant, the authors are not aware of any other published data relating to maintenance failure probabilities' (Williams and Wiley, 1985).[16]

Such data as are available are subject to the normal restrictions on information in the areas we have been examining. 'There is little systematic recording in industry of human errors, the contributory work-place factors and the error consequences. Also, where records do exist, they are frequently confidential to the company involved' (Human Reliability Assessment Group, 1985). Finally, one expert in the area believes the difficulty in obtaining these types of data relates to the doubtful usefulness of the category of human error:

The nature of the tasks in modern systems, being related to supervisory control, involving problem-solving and decision-making in which adaptation to unfamiliar situations is crucial, makes it very doubtful whether a category of behaviour called errors can be meaningfully maintained and, consequently, whether error data can be collected for reliability prediction. (Rasmussen, 1985, p. 1187)

In the Institute example, we have already seen how safety features can increase risk. In general, the attempts of safety engineers to increase safety by increased use of redundant or back-up systems can also increase hazards. When something goes wrong with one system or part of a system, it does not always get a high priority for repair and maintenance, as there is already a back-up system in operation. Rausch (1986, pp. 14–15) cites an accident in which there were seven critical instrument failures. He questions, 'Would the addition of another safety device have prevented the incident? I doubt it. If you don't maintain one instrument, is there any better chance of maintaining two, three or four?' (It's those damn workers, again, shirking maintenance.)

The basic problem is that accidents are made inevitable by the complexity of the technology, the expansion of the industry and management control over the work process. Kletz's conclusions on Bhopal are relevant here: 'The Bhopal incident [sic] illustrates the general point that, whatever is spent on safety devices, they can go wrong.' Increased complexity provides more possibilities for mistakes to be made or things to go wrong: 'Multiples of sometimes small failures, making unexpected connections and moving along unplanned or hidden paths, may set off a large and catastrophic accident, leaving the managers uncertain about how or where to intervene' (Engler, 1985, p. 497).

We are now approaching the heart of the matter. Recently, from an analysis of high-risk systems such as nuclear power, air and sea transport, chemical plants and space exploration, Perrow has argued for the existence of what he calls 'normal accidents'. These may also be termed system accidents, as they are a rare but inevitable outcome of the structure of complex systems. Thus highly complex systems run the risk of strange, unexpected and unforeseen interactions between various parts of the system, whose synergistic effect will be greater than the individual interactions themselves:

Systems that transform explosive or toxic raw materials or that exist in hostile environments appear to require designs that entail a great many interactions which are not visible and in expected production sequence. Since nothing is perfect – neither designs, equipment, operating procedures, operators, materials and supplies, nor the environment – there will be failures. If the complex interactions defeat designed-in safety devices or go around them, there will be failures that are unexpected and incomprehensible. If the system is also tightly coupled, leaving little time for recovery from failure, little slack in resources or fortuitous safety devices, then the failure cannot be limited to parts or units, but will bring down subsystems or systems. These accidents then are caused initially by component failures, but become accidents rather than incidents because of the nature of the system itself; they are system accidents, and are inevitable, or 'normal' for these systems. (Perrow, 1984, p. 330)

One example Perrow cites of complex system interactions leading to an accident is the 'common-mode failure', when a failure of one component affects two modes of operation in the system. This failure has become most familiar from the nuclear industry.

Hagen [editor of Nuclear Safety*] concludes that potential common-mode failures are 'the result of adding complexity to systems design'. Ironically, in many cases, the complexity is added to reduce common-mode failures. The addition of redundant components has been the main line of defence, but, as Hagen illustrates, also the main source of the failures. 'To date, all* *proposed "fixes" are for more of the same – more components and more complexity in system design.' The Rasmussen safety study relied upon a 'PRA' (probabilistic risk assessment), finding that core-melts and the like were virtually impossible. Hagen notes (p. 191) that PRAs, 'using established techniques of reliability and statistical analysis', constitute the main source of public reassurance about such potential dangers. But, he notes, this involves a 'narrow definition' of common-mode failures, and the analysts 'are busily working in an area which has not been shown to be the main problem'. The main problem is complexity itself, Hagen argues. With that we can agree. (Perrow, 1984, p. 73)*

Perrow's general analysis is confirmed by noted industry analysts: 'The typical accident in an installation of balanced design is not caused by only one simple equipment fault or human error; on the contrary, major events will depend on a complex chain of events including equipment faults, latent risky conditions from repair and modifications as well as human mistakes and decision errors' (Rasmussen, 1982, p. 144). Rasmussen cites another analysis which found 'major rare events are probably dependent on five to ten faults and abnormal conditions . . . An accident is typically caused by a sneak path for events, created by the accidental timing of a considerable number of normal and erroneous human acts together with latent risky conditions and equipment failures' (p. 153). The most important thing about these sneak paths is that they cannot all be predicted by risk analysis and assessment.

Perrow argues:

Our major concern is with systems that appear to be irretrievably complex, or at least will remain so for the next few decades. Generally these are systems that transform their raw materials rather than fabricate or assemble them . . . Most of these are quite new, but it is significant that chemical processing is not. While experience has helped reduce accidents, accidents continue to plague transformation processes that are fifty years old. These are processes that can be described, but not really understood. They were often discovered through trial and error, and what passes for understanding is really only a description of something that works. Some of industrial chemical processing is of this nature. (1984, p. 85)

Perrow presents the ammonia industry as an example of one such area of concern.

A very mature operation has to be shut down for repairs about 30 percent more frequently than expected, generally from major equipment failures, and its plants have a fire every eleven months. If the problems of the ammonia industry have not been 'wrung out' of the system by now, it may be that they are endemic to the system – normal accidents. (1984, p. 105)

The increasing complexity of plants increases the possible number of system failures. The increasing number of plants increases the number of locations at which such failures are possible. 'Furthermore, as design complexity increases, it will become more difficult to identify all potential failure modes, although we can expect a parallel improvement in techniques. Even when the disaster risk from individual plants is acceptably low, the cumulative risk from a large number of plants is correspondingly higher' (Wood, 1985). With hazardous processes 'the typically accepted risk is now one in a million per year', Kristin Shrader-Frachette told *Business Week* (24 December, 1984 p. 51).[17] But, because of the rising number of possible events, 'the aggregate probability of something catastrophic happening is much higher than people realize'. Risk analysts admit that their calculations may be off by a factor of ten or a hundred. Thus the 'acceptable level of risk – one in a million – may be one in a trillion or one in 10,000'. There are 15,000 chemical plants in the USA, for example.

Since capital cannot guarantee safety, it has to persuade people accidents are normal or acceptable or else prove emergency-response systems are well developed. This is where the whole risk 'debate' arises and where the comparisons with Arpège or tear-gas (different strokes for different folks) are trotted out.

Risk assessment has been adequately debunked elsewhere.[18] Its major fault is that it is class-blind: it ignores the class distribution of risks. As EPA administrator Lee Thomas has so well encapsulated it, it is in the nature of things that certain people face more risk than others. Yet there is nothing natural about the fact that those who face the immediate risks of chemical production are generally poor, working class and marginal. Chemical executives do not generally live next door to their plants.[19] The plant-location policies of chemical companies are no more natural than the products they produce. Risk may be seen as a mystification which attempts to hide the reality of risk as a class relation, another example of the power of capital over our lives.

The industry recognizes also that accidents are inevitable. Thus UCC was criticized by industry sources for saying the Institute plant was safe. In recognizing this, the industry is moving more towards crisis communications, crisis management and evacuation planning. At the same time it increases reassurance operations, beefs up its risk analyses and induces the wider community to share not only in the experience of risk, but also in its management.[20]

We also must recognize the inevitability of accidents. This book also should be seen as a type of crisis communication. We naturally take an oppositional stance to the industry. Perrow's general conclusion is apt here: '. . . the problems are not with individual motives, individual errors, or even political ideologies. The signals come from systems, technological and economic. They are systems that élites have constructed, and thus can be changed or abandoned' (Perrow, 1984, p. 352).

9 WE ALL LIVE IN BHOPAL

This chapter describes various struggles over the safety of toxic industry in the period following Bhopal. These encompassed both peripheral and metropolitan countries, both community and trade-union struggles. The blocking of plants at Livingston New Town and Phuket are recounted as successful examples of the strategy of rejection. In the first example, a Scottish town rejected an industrial gases plant proposed by UCC. In Phuket a peripheral community took extreme direct action to block the proposed establishment of a metal-refining plant. At Geismar a trade union used the question of safety as part of a traditional labour dispute. The Basle example provides even richer fare, as it brought Bhopal back home to Europe, just as Institute did in the USA.

LIVINGSTON NEW TOWN

In February 1985, UCC was denied permission to set up a plant intended to mix gases for the Scottish electronics industry in a town with a 26 percent unemployment rate. The $66 million plant had originally been given red-carpet treatment when it was announced the previous summer. Following Bhopal, UCC's planning permission was frozen until the public, the

Health and Safety Executive and the local district council were consulted. UCC distributed slick promotional brochures to thousands of people in the vicinity of the Deans area, where the plant was to be located.

Among the toxic chemicals to be used at the plant were arsine and phosphine. The known toxicity of the gases, combined with the recent example of Bhopal, aroused public opposition to the project. In response, UCC distributed reassuring pamphlets and commissioned a consultant's report, which calculated the risk of local residents being affected by a toxic-gas leak to be less than the risk of being struck by lightning – around one in 12.5 million.

This ploy did not wash. UCC was denied permission on 10 February 1985 after two packed angry public meetings at which residents rejected UCC's attempts to reassure them. The tone of the meetings was indicated by a local protester who was quoted (*Financial Times*, 8 February 1985, p. 2) as saying, 'We don't want you here. Go away once and for all.'

The defeat was a major blow to UCC. While industrial gases represented only 15 percent of UCC sales in 1983, they contributed 30 percent of operating profits. UCC's approach was partly responsible for the defeat. A more open approach might have persuaded the Livingston New Town development corporation to defer the decision until the Health and Safety Executive examination of UCC's safety case for the plant. One major mistake UCC made was to refuse to talk about Bhopal, on the grounds that it was not relevant to Livingston, as MIC would not be among the chemicals used there. Similarly UCC's approach was strongly criticized by local residents. 'The Union Carbide men just could not relate to the people', was a typical comment after the meeting. 'They did not send in a powerful enough team and showed slides that no one could understand' (*Financial Times*, 8 February 1985, p. 8).

Phuket

If the Livingston refusal was a surprise, it had none of the shock impact of 23 June 1986 when residents of Phuket, Thailand, burned down a tantalum plant. The attack on the plant was attributed variously to outside agitators, political conflicts and commercial rivalry. Observers who accept such explanations ask why the opposition became so strong in May 1986, when the forthcoming general election was announced and some 80 percent of the plant had been built. In fact, opposition began when construction started the previous year. The reason for the late development of the mass protest has been given in an editorial in the *Nation* (20 June 1986): 'What has become clear is, the residents of Phuket were unaware of the plant's purpose. They were uninformed about tantalum, how it is refined and how wastes from such an operation would affect their well-being as well as how they would be handled.'

At first the Thai government had strongly supported the project when it was proposed in 1981. The plant was intended to refine tantalum from tin slag. Tantalum is a heat-resistant metal used in computer components, nuclear reactors and missile warheads. Investment in the plant was either $44 million or $75 million: sources differ. The consortium involved, TTI Corp. (Thailand Tantalum Industry Corporation), was formed mainly by local Thai companies and investors. Some 20 percent of the capital invested was either Malaysian or from the World Bank's International Finance Corporation. Multinational capital was therefore only marginally involved.

Opponents of the plant were worried about the plant's use of such toxic chemicals as hydrofluoric acid and thorium dioxide and its disposal of radioactive waste. Protesters were particularly worried about the plant's effects on the area's tourism industry, the main source of local employment. Opposition began with the construction work. By December 1985 TTI began to feel growing pressure from the protest. TTI held a meeting to assure village heads and chiefs of the plant's safety. The opposition continued to grow, however. *Asiaweek* (6 July 1986, p. 24) reported: 'The local population, it seems, feared the prospect of an industrial disaster like Bhopal or Chernobyl. According to some reports, videos suggesting such scenarios had been circulating in the town.' In April TTI claimed that the protest was politically motivated. TTI responded to the growing protest by launching a newspaper advertising campaign promising no pollutants or toxic substances would be discharged from the plant.

On 1 June 1986, some 50,000 people demonstrated in Phuket against the plant's proposed opening the following August. A petition signed by 64,800 people threatened a boycott of the Thai general elections due the following month if the government did not stop the plant by 6 July. On 9 June the Phuket municipality council, in a special session attended by about a thousand Phuket citizens, decided to notify the government formally of its opposition to the plant. In response to the protest, the government claimed it had taken steps to ensure safety and no pollution. These steps included provision of an alarm system for the store of hydrofluoric acid used in the refining process. The Office of Atomic Energy was to check radioactivity levels of plant waste. As protests escalated, the response changed. On 15 June, TTI announced the opening of the plant would be postponed by six months. A board member said, however, that relocation of the plant 'is an impossible proposal' on grounds of cost.

The government response changed also. They undertook a fact-finding exercise, which they hoped to have completed by 2 July. In response to this, the chairperson of the Phuket environmental control committee warned that a massive demonstration would be held on 6 July if the government did not announce its decision by 2 July. The visit by a government minister on 26 June to meet local people to discuss the issue was the cause of the massive demonstration that finally resulted in the plant's destruction. That the protest was a popular one cannot be denied. While the people of Bhopal hung effigies of Warren Anderson after the killing, the people of Phuket hung effigies of major investors in the plant from lamp-posts during their campaign against the plant. One candidate's election posters were vandalized as he was believed to be an investor in TTI. Other residents began boycotting Coke because the local bottler, Thai Pure Drinks, held shares in TTI. By 22 June, some eighteen local groups supported the campaign. Protesters obtained and used information from residents of a German town where a similar tantalum plant was located. Officials of the Tourism Authority of Thailand opposed the plant because of its adverse effects on the tourism industry.

Massive crowds of people waited from early in the morning to meet the government minister at a public hall. The long wait combined with hunger and the midday heat made the crowds restive. The crowd refused to disperse when called on to do so: instead they attacked the hall. Part of the crowd went to the Phuket Merlin Hotel, where the minister was reported to be. When the minister did not leave the hotel to meet the crowd, they attacked the hotel. Some thirty police did not intervene to protect the hotel. The owners of the hotel were also investors in TTI. Crowds also attacked three banks which had invested in TTI. Residents were reported to have withdrawn their deposits before the banks were attacked.

Following the attack on the hotel, residents attacked and burnt the tantalum plant. The police guarding the plant had been withdrawn before the attack. 'The reason given was that the presence of police guards there might cause a misunderstanding among the people, who might think that the government was protecting the plant. Occasional police patrols were ordered instead' (*Bangkok Post*, 28 June 1986). No one was injured in the riots, though police fired tear-gas to quell the crowds. A state of emergency was declared and the government prepared to deploy troops to restore order. Some fifty people were arrested. When they had not been released by the following weekend, there was talk of further demonstrations in Phuket.

The government's account emphasized that the riot was caused by outsiders. The explosion of the popular opposition to the plant into the fiery protest of 26 June was attributed to political manipulation and outside agitators. The Prime Minister of Thailand, Mr Prem Tinsulanonda, was reported as saying the violence was not caused by Phuket people. The following day, police said two suspects arrested the previous night had admitted they had been paid to create unrest. The *Nation* (28 June 1986, p. 8) reported: 'Several suspects have told police they had been paid to cause trouble during the riots. However, police have yet to produce the alleged master-minds.' Eight local protest leaders met the Fourth Army Region commander on 27 June and assured him they had no part in the violence, for which they blamed the local governor.

On 27 June, TTI announced it would abandon the Phuket plant as beyond repair. It said it would explore possibilities of an alternative site, but only if it was guaranteed government support. This was not forthcoming, though the government had agreed to assist the owners in the matter of compensation. Disagreement arose between TTI and its insurers over the extent of damage. The plant's insurer, Bangkok Insurance Co., said damage amounted to 100 million baht, while a TTI engineer estimated it at 2,000 million baht. According to government inspectors, more than 60 percent of the plant was beyond repair. The Bangkok Insurance Co. refused to pay indemnity as it said TTI had not bought insurance cover for riots, strikes and civil disorder, though it had urged

TTI to purchase such insurance. TTI denied this, saying it had specifically sought such coverage but it had not been provided.

The governor of the Tourism Authority of Thailand was reported as saying the riots resulted from Phuket people's intention to protect the Ko Phuket island's environment and natural resources, unlike foreign riots: 'It was not like a war or revolution caused by political conflicts; it was due to their good intentions.' This conciliatory line was sustained by the government, despite accusations that it had failed to deal sharply with 'mob rule'. The government was concerned also for its own political popularity: its ascription of the actual riots to outside elements could not blind it to the depth of popular opposition to the plant. Whatever the origins of the violence, the popular rejection of the tantalum plant in Phuket must be considered a major victory for the South-East Asian environmental movement. The Phuket incident also shows how successful direct action by a united community can be.

Geismar: Bhopal on the Bayou?

The dispute between the US Oil, Chemical and Atomic Workers' Union (OCAW) and BASF Wyandotte Corporation at BASF's Geismar, Louisiana, plant is the first major example of a labour–management dispute in which both sides have used the spectre of Bhopal to shore up their positions. The dispute is important in that it reveals differing interpretations of Bhopal by chemical management and unions. While BASF stresses sabotage, the unions stress management practices.

The dispute began in June 1984 when around 370 members of OCAW's

OCAW's Local 4-620 erected this hoarding on the interstate road near Geismar.

Local 4-620 were locked out by BASF after management and union were unable to agree on a new contract. At a news conference in Washington in March 1986, OCAW charged BASF with 'inviting another Bhopal disaster' at the Geismar plant. Since the lock-out the Union charged, 'BASF has been running a complex chemical manufacturing operation with inadequately trained contract workers who are violating critical safety procedures on a regular basis.' BASF has reacted to these allegations of inadequate safety by presenting the unionized workers themselves as a possible threat to safety.

Drop in Chemical Union Membership
(Chemical Workers Only)

Union	1980 Membership	1983 Membership
Oil, Chemical and Atomic Workers' Union	80,000	65,000
International Chemical Workers' Union	65,000	55,000
United Rubber Workers	180,000	130,000
United Steelworkers	60,000	52,000

The background to the dispute is the drop in union membership in the chemical industry over the last decade (see table). In the case of OCAW, overall membership dropped from 156,000 in 1980 to 110,500 in 1986. This drop in membership results not only from the general expulsion of labour from the chemical industry over the last decade,[1] but also from corporate attempts to expel unionized labour. OCAW has claimed BASF has a policy of union-breaking: in 1979, according to the union, more than 2,000 BASF workers were union members; by mid-1985 only 200 members were under union contract with BASF. Before the Geismar contract ran out, BASF told the union's local in July 1983 they were no longer allowed to 'consult with aggrieved employees relative to contractual violations'. BASF forced the union to move from its on-site offices. OCAW's appeal to the US National Labor Relations Board resulted in a ruling that ordered the company to reinstate the union and the contract. When discussions broke down over a new contract, BASF locked out the union members in June 1984.

The basic cause of the dispute was management's attempts to cut costs by using unorganized labour. A BASF spokesperson, Leslie J. Story, estimated that by using contract workers who are paid $5 less per hour than union members, the Geismar plant could save $2 million a year. In its negotiations, BASF were seeking concessions on wages, health-care, worker seniority and subcontracting. The union claims BASF was looking for an excuse to lock out its members and that its negotiations prior to the lock-out were not in good faith. After the lock-out, BASF's position hardened even more. One company offer for a return to work, which included offering 200 workers amounts from $6,500 to $10,000 to resign, seems to indicate a desire to be rid of union labour entirely and to replace trade-union members with contract workers at will. According to OCAW (Summary document, 4 April 1986):

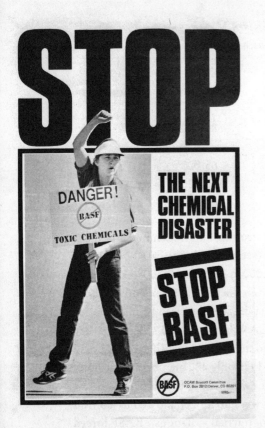

> Currently, BASF is demanding that 181 union members resign and permit themselves to be permanently replaced with temporary workers. They are also insisting that the union allow the returning 194 workers to be subject to disqualification for any reason, including inability to relearn the job, health and hostility towards the company. They further insist that the union allow them to replace any worker at any time with temporary labour.

In response to the lock-out, OCAW waged an imaginative corporate campaign against BASF. A major plank in that campaign was that the continued operation of the Geismar plant without organized labour represented a threat to safety. Union officials presented unionized workers as a guarantee to the community against unsafe practices by BASF: workers under contract have the right to refuse unsafe work, thus preventing the company from cutting costs by ignoring safety procedures. Further BASF use of contract workers meant the corporation had no direct control over safety regarding these workers, nor would the contract workers be as experienced in dealing with toxic chemicals as the union members were. Esnard F. Gremillion, chairperson of the union local, told *Industry Week* (3 February 1986, p. 17), 'It's particularly worrisome to have chemical firms sidestepping their responsibility for health and safety by farming out work'. The union's Richard Miller has said that BASF 'might try to shrug off liability for a chemical accident', should one occur, on the basis that it had no direct control over its subcontractors' safety programmes.

In March 1986 OCAW published a report which, using internal BASF documents, found that the number of workers exposed to toxic chemicals jumped from one during the two-month period prior to the lock-out to twenty-six for the two months after the lock-out. Similarly the number of safety violations, 'tag-out violations' which could permit the escape of toxic fumes, jumped from nine to forty. The union said these illustrate an apparent lack of familiarity with safety procedures inside this complex and dangerous plant. The union report also included a May 1985 BASF report which listed toxic chemical releases – phosgene (three), ethylene oxide (one), toluene di-isocyanate (one) and accidents – chlorine (five) and hydrochloric acid (one): these were described as 'selected incidents'. The union also pointed out an increase in the labour turnover rate from 0.5 percent for union members to 38 percent for one subcontractor after the lock-out.

The union further charged that many of the contract workers' background was in construction, and that therefore they were not experienced enough to work in a highly complex petrochemical plant. BASF responded by saying that contract workers had almost as much experience of chemical plants as had the OCAW members whom the corporation had locked out. The union responded to this by claiming that even those contract workers with general chemical-plant maintenance experience hadn't as much experience in producing the specific, highly toxic chemicals produced in Geismar as union members had.

BASF responded to the union campaign by launching a publicity campaign of its own. It hired New York public-relations firm Hill & Knowlton to counter the union campaign. Taking up the issues of Bhopal and safety, Leslie J. Story, general manager at Geismar, told the *New York Times* (26 May 1986) that BASF ordered the lock-out because it could not afford to operate the plant with disgruntled workers who might commit sabotage. BASF claimed that because of the toxicity of the chemicals involved, it was not possible to allow workers in dispute with the company to operate the plant: 'The potential danger posed by carelessness, negligence, walk-outs and sabotage is simply too great to justify leaving the work-force in place' (quoted in *Multinational Monitor*, 15 March 1986, p. 10). BASF's response included an advertising campaign in the local press and on television. The union report quotes one company ad claiming that the employees at BASF 'just completed 4 million work-hours without a lost-time injury. That means working 4 million hours without an injury serious enough to require the loss of time from work.' On this basis, BASF awarded its Geismar employees the Triple Crown of Safety, an award that was advertised in a local Baton Rouge paper, the *Sunday Advocate*. This award seemed less impressive when the company admitted that the 4-million-hour safety record excluded the contract workers, as they did not work directly for BASF. Recipients of the award included clerical, laboratory, instrument and management workers, those least likely to be exposed to the hazards of the plant's operation.

The union continued its campaign to throw doubt on the plant's safety, with a major use of internal company papers to which it seemed to have almost magical access. The union made public a BASF memo on a release of the nerve gas phosgene on 2 April 1986, which quoted a company official, 'I have seen evidence of a deterioration in commitment [to safety] lately.' This release of phosgene became a source of controversy. The union produced a BASF memo which said 75 lb of phosgene leaked that day, seven people sought medical attention and traces of the chemical were found near the Mississippi river, outside the plant. Earlier BASF had said 50 lb were involved, no one at the plant was exposed and the chemical did not escape from the plant site. BASF later agreed that six workers were exposed but denied any of them were injured.

The union also threw doubt on other safety assurances from the company. One union publication (*Basagram*, 1985, vol. 7, p. 4) gave an example of the company's style of safety assurances: 'If a group of people were holding an important meeting in a closed room at normal room temperature and there was an open drum of TDI [toluene di-isocyanate, a chemical relative of MIC] in the room, people could finish their discussion, but for comfort they would be better off to finish the meeting outside the room',

Parsippany, New Jersey, 3 June 1987:
Demonstrators blockade the entrance to BASF's US headquarters during an hour-long 'lock-in' of company executives who had locked out 370 union members at the Geismar, Louisiana, plant since 1984.

while 'if a truck spilled TDI, only those immediately adjacent to the spill would need to evacuate'. The union then reprinted a newspaper report from Springville, Alabama, of a spillage of 2,000 gallons of toluene diisocyanate on 9 March 1986 in which eighteen people were injured, twenty miles of highway were closed and the entire population (1,500 people) of Springville were evacuated.

The union campaign against BASF involved it in some unusual alliances. Contact with the BASF chemical workers' union in West Germany was traditional enough, even if the union was less successful than it hoped: it received $10,000 and a letter of solidarity addressed to its 'colleagues' in Geismar. Less traditional was its contact with the West German Green Party, from which the union received more support than from its brother union. In April 1986, two West German Green members of parliament, part of a visiting delegation of what were hilariously described as 'German officials' in a local paper, were refused permission to visit the Geismar plant to investigate the union charges. The union also lodged a complaint with the Organization for Economic Co-operation and Development, maintaining BASF's policies broke the Organization's guidelines on multinational enterprises. BASF's response was reported to be that it was not a multinational and thus was not subject to the guidelines!

At the time of writing there had been no resolution of the dispute. The union campaign co-ordinator had said there is more involved in Geismar than simply a labour dispute, and even if the Geismar issues were settled tomorrow, 'there are certain things set in motion that we wouldn't want to stop'. These certain things include the Congressional investigation on safety and the possibility of further union-environmentalist alliances.

'Chernobasle': Bringing Bhopal Home to Europe

'They said it couldn't happen here. They said we are protected. The people felt safe. But the fact of the matter is – it did happen here.' – Swiss woman (quoted in Chemistry and Industry, *1 December 1986, p. 802)*

'Seveso and Bhopal seemed so far away. This [Sandoz fire] wasn't.' – Hans Kaufman, Zürich financial analyst (quoted in Wall Street Journal Europe, *13 November 1986, p. 4)*

The fire at Sandoz's Warehouse 956 in its chemicals complex at Muttenz, Basle, in Switzerland at the beginning of November 1986 brought the spectre of Bhopal to the heartland of the Western European chemical industry. The West German Social Democratic Party chairperson, Willy Brandt, described the fire and resulting pollution as a 'Bhopal on the Rhine' (*Guardian*, 15 November 1986, p. 8). 'Chernobasle', as it was quickly christened by the youth of Basle, performed the same role in Europe that the Institute leak had performed in the USA. It raised questions regarding the availability of information, government and industry responses to chemical accidents, inadequate warning systems, and government inspection and regulation of the industry; it also raised general questions about the industry's safety standards and control of information. For the industry, Chernobasle and the leaks over the following weeks from chemical plants in West Germany and Switzerland were a public-relations disaster. The industry feeling was reported to be: 'If this can happen in Switzerland and Germany, it can happen anywhere' (*Business Europe*, 1 December 1986, p. 1).

The accident began with a fire at Sandoz's warehouse in Muttenz. The fire released a ten-mile-long toxic cloud of foul-smelling gas. All of Basle's residents were told to remain indoors, many schools were closed, and all traffic into the city was briefly halted. It took six hours to put out the fire. Water used by the fire-fighters carried toxic chemicals into the Rhine river, leading to a virtual killing of all life in the nearest stretch of the river, including the death of such hardy, pollution-resistant species as eels. The Dutch transport and public works minister said that one year's normal content of mercury had been dumped into the Rhine in one day.

While this seemed bad enough, others claimed Chernobasle was a near-Bhopal. Phosgene was reported to have been stored in a tank 250 yards from the fire. An expert back-bench member of parliament for the West German Social Democratic Party said the Sandoz plant also processes organo-phosphate pesticides, whose effects are similar to those of nerve gas. 'If the twelve tonnes of those organo-phosphates stored in Muttenz had gone up, the result would have put Bhopal in the shade' (*Observer*, 16 November 1986, p. 15). The fire also released dioxin. A 'very low concentration' of dioxin was found by Sandoz in the remains of Warehouse 956.

The fire at Muttenz undermined the industry's claims that in metropolitan Europe it was so well prepared and its safety measures were so good that no accident similar to Bhopal could occur. It quickly became obvious that Sandoz was ill-prepared for a fire in its storage area. Sandoz admitted it had concentrated its safety spending on its manufacturing areas. It did not expect an accident to occur in its storage facilities. It also expected fires to be civilized and to occur during working hours only: 'We believed that a fire could only occur there during working hours and we drew up our safety precautions accordingly', Sandoz executive board member Hans Winkler told *Business Europe* (1 December 1986, p. 1). Alas, the fire worked overtime by starting at midnight. The special fire-fighting teams of the Basle chemical companies were reported to be off duty in the evening. The warehouse involved was previously a machinery store and was not equipped with sprinklers.

Similarities with Bhopal abounded. Local doctors were not informed of what substances had leaked and so found it difficult to treat those exposed: 'We had no idea of what we were dealing with', one local doctor was quoted as saying. There was immediate controversy over the toxic cloud's health effects. While the Swiss authorities quickly claimed that the toxic threat had passed, one Basle doctor told a Greenpeace press conference that as many as two-thirds of the population of Basle might be affected: the doctor cited the fact that two-thirds of sixty-eight patients of his were complaining of headaches and breathing or throat disorders days after the accident. Physicians for Social Responsibility issued a survey claiming up to 70 percent of the population suffered from such effects as headaches, eye and throat irritation, nausea, fever and vomiting.

Another similarity related to the de-staffing policies of chemical companies. A local trade-union official told the *Guardian* (21 November 1986, p. 8) that Sandoz had cut back staff by 15 percent and this increased the risk of accidents. Because of Sandoz's rationalization plans, which had boosted profits, 'many people are stretched to work at full capacity all the time and that raises the accident risk'.

Chernobasle also illustrated the by now common problems with warning systems. It took the Swiss authorities forty hours to sound the alarm about the pollution of the Rhine internationally. It was explained that a breakdown had occurred in the early warning system intended to warn countries along the Rhine of chemical accidents. The EEC Commissioner for the Environment, Stanley Clinton Davies, said Swiss warnings to neighbouring states were 'grotesquely inadequate'. The French Minister for the Environment claimed that officials at the French town of St-Louis, which faces Basle across the Rhine, were not alerted until long after the fire began.

Similarly Chernobasle undermined industry claims that more stringent regulations in metropolitan countries ensured Bhopal-type accidents could not occur. The Swiss President admitted that the Sandoz case proved the 'existence of good laws' does not protect the public from environmental catastrophes. Sandoz's storage area had been inspected four days before the fire by a Swiss Fire Prevention Service official, who concluded the area was safe. The insurance companies were less sanguine. A 1981 report by the Zürich Versicherung insurance company was reported to have said safety precautions were totally inadequate. The report, leaked to the West German Green Party, said the probability of a fire was 'medium', with 'great' damage likely in such a case. The company was reported (*Observer*, 16 November 1986, p. 15) to have refused to cover Sandoz for third-party liability following an accident. Sandoz responded that it hadn't taken a third-party policy with the Swiss company on the basis of cost alone: it had been offered a cheaper policy by a West German company.

Sandoz's Response

Sandoz's response was unsatisfactory in various ways. In certain ways it resembled UCC's response to Bhopal; in other ways it was deeply different. Sandoz was much less forthcoming than UCC in making itself publicly available. It didn't hold its first press conference until a week after the leak. Its president didn't appear in public for three weeks. It did not make an official list of the warehouse's contents publicly available until four days after the fire. This official list involved a change in Sandoz's estimate of the amount of chemicals involved in the fire: the initial estimate of 600–900 tonnes rose to 1,300 tonnes, including 934 tonnes of pesticides and 12 tonnes of mercury-containing compounds. This list showed that some 956 kg of the warehouse's total contents of 6,653 kg of chemicals was not on the warehouse's inventory. Other contradictions also came to notice. At the company's first press conference it distributed a written statement by the Zürich Versicherung insurance company stating that it had not passed on to Sandoz the 1981 risk-analysis report warning of the danger of fire and environmental damage if sprinklers were not installed in the warehouse. This contradicted previous Sandoz statements that it had been orally informed of the report's contents.

Sandoz also stressed that it could not rule out arson as a cause of the fire, though the local police were reported to be unimpressed by a phone call claiming responsibility for the fire for the Red Army Faction. This line was abandoned by Sandoz when the cause of the fire was finally reported. The cause of the fire brings us back to the level of candles causing nuclear-reactor accidents: 'Sandoz . . . said that the fire at its warehouse was caused by an animal gnawing on wiring in the building' (*New Scientist*, 27 November 1986, p. 17).

Nevertheless Sandoz strode out there to take its moral responsibilities seriously in public. 'We're deeply moved by the extent of the accident. We take the full *moral* responsibility,' said Hans Peter Sigg, executive board member of Sandoz (*Wall Street Journal Europe*, 14 November 1986, p. 4), following UCC's line after Bhopal in stressing moral over legal responsibility. As for its legal liability, Sandoz formally pledged compensation for 'proven claims' of damage caused by the fire. Another Sandoz spokesperson, M. Fazel, said: 'It goes without saying that damages were caused by this fire. We are responsible, and there is not a shadow of a doubt that we will take care of it . . . we feel badly about it, and I would say we are shocked and saddened by what has happened to us' (*Chemical Marketing Reporter*, 17 November 1986, p. 69).

For those more worried about the material than the moral aspects of the crisis, Sandoz's technical director explained the company had liability insurance 'of a level that is normal for the [chemical] industry' (*Irish Times*, 14 November 1986, p. 8). This was estimated to be $58–88 million by European observers and over $100 million by *Business Week*.

Nevertheless, Sandoz shares were affected, as were shares in the other two major Swiss chemical companies: Sandoz stock lost approximately 16 percent in value, while Ciba-Geigy fell 5 percent and Hoffman LaRoche 3.5 percent.

Interestingly, Sandoz was attacked in the pharmaceutical industry press for its complacent attitude to safety. H. Schwartz, columnist with the industry trade paper *Scrip*, stressed the problems of arson and sabotage as the next form of possible attack on the beleaguered drug industry but also slated the complacency of company officials: 'In Basle, the complacency was that of company officials charged with the safety and anti-pollution defences of the plant, who had apparently never thought through what a warehouse fire fought by a vast amount of water could produce' (*Scrip*, 15 December 1986). Anxious to protect the good image of the drug industry, Schwartz suggested that drug companies should get out of poison and pesticide production, or at least should spin these activities off to separate companies.

As a general strategic response to the fire, Sandoz announced it intended increasing its present shift away from synthetic chemicals to synthetic biologicals. As intermediate measures Sandoz proposed a general reassessment of safety, an increase in security guards and in storage space. It is also following the US chemical industry lead in reducing its storage of hazardous chemicals. Sandoz also developed a new storage concept, requiring more space between storage tanks holding flammable materials. They admitted they did not have enough storage space in Basle for this. Therefore they intended to buy in some of the active products involved from other companies: this would increase costs as would moving some hazardous production out of Basle. In an attempt to regain public confidence, Sandoz also announced the setting up of an independent expert commission to produce a report on security and environmental-safety questions. This report was to be made available to the industry.

The popular response to the Chernobasle disaster indicates the kind of reception that industry reassurances are likely to receive in the future. In France an environmental coalition called for a boycott of all Sandoz products. On Sunday, 14 December 1986, more than 30,000 people took part in protests against the pollution of the Rhine. Groups of protesters took control of bridges over the Rhine in Bonn and other towns along the Rhine, carrying banners attacking chemical companies and blocking traffic. Demonstrations crossed national borders, as does the chemical threat: demonstrations took place in Switzerland, West Germany and the Netherlands.

The reaction of the people of Basle also deserves mention. On 8 November 1986, 10,000 people demonstrated in Basle. Most demonstrators were schoolchildren, angry at the way their schools had been quickly reopened as part of a reassurance campaign that all was well. Banner slogans included: 'We do not want to be the fish of tomorrow', 'The fish are helpless, we are not' and 'Bla . . . bla . . . blabla . . . bla'. The following day protesters broke up a public meeting attended by Sandoz and government officials to reassure people the crisis was past. The reassurers were sprinkled with Rhine water and pelted with dead eels.

Chernobasle has resulted in continuing hostile reactions to minor spills in Switzerland. 'Before the Sandoz disaster, people accepted small accidents like the Ciba-Geigy one, just like British miners accept the risks of their trade, but it does not stop son following father down the pit,' a trade-union official told the *Guardian* (21 November 1986, p. 8). 'But now everyone is scared about minor incidents.' More interesting yet in showing public perceptions of the chemical industry's behaviour was a Swiss television poll taken three weeks after the Sandoz fire. Less than a quarter of the residents in the Basle region believe Sandoz will actually improve safety. As almost half the jobs in and around Basle are provided by the chemical industry, the local residents are in close contact with the reality in the industry and must have strong grounds for their suspicions.

Chernobasle's Knock-on Effect

More general criticism of the industry was inescapable. It wasn't possible to dismiss Chernobasle as a rare, chance occurrence after two further large accidents occurred that week at chemical plants. An explosion at the Devnia PVC (polyvinyl chloride) plant in Bulgaria killed seventeen people and injured nineteen, many seriously. The same week, an accident at a Transquímica chemical plant in Mexico City killed six people. Nor were they rare in the area. The mayor of St-Louis, the neighbouring French town, complained that the fire at Sandoz was the third accident that year. In June and September 1986 bromine had leaked from Ciba-Geigy's Basle plant and in September 1986 a fire broke out at Sandoz's plant at Huningue, on the French side of the border. This was all bad enough. Even worse for the industry, the knock-on effect also focused attention on its normal discharges and emissions.

> Toxic Accidents on the Rhine, November–December 1986
>
> **1 November:** 'Chernobasle'.
> **5 November:** Ciba-Geigy spills Atrazine into Rhine.
> **6 November:** Lonza subsidiary closes sections of Waldshut plant after emissions.
> **7 November:** Second spill of polluted water from Sandoz Muttenz site.
> **20 November:** Phenol emission from Ciba-Geigy resins plant.
> **21 November:** Leak of toxic herbicide 2,4-dichlorophenoxy acetic acid into the Rhine from BASF plant. Hoffman LaRoche subsidiary released methyl vinyl ketone into Rhine.
> **22 November:** Hoechst releases chlorinated hydrocarbons into a tributary of the Rhine.
> **25 November:** Bayer discharges 200 kg of a chloride disinfectant into the Rhine.
> **26 November:** Bayer discharges 800 kg of methanol alcohol into the Rhine.
> **28 November:** BASF spill of ethylene glycol into Rhine.
> **1 December:** Fifty litres of highly inflammable liquid gas escape from factory at Prattelm on the Rhine into a sewer system.
> **2 December:** Five cubic metres of PVC (polyvinyl chloride) discharged by Lonza into the Rhine near Waldhut, West Germany.

Unfortunately for the industry, Basle was followed by a string of further chemical spills and leaks (see box). Some of these spills would have previously excited little media coverage: 'Chemical industry experts contend that spills such as the BASF leak [see box] occur often but go unnoticed when public attention isn't focused on the industry' (*Wall Street Journal Europe*, 24 November 1986, p. 4). With media attention now firmly focused on the industry, these leaks proved to be a major embarrassment. Further problems arose when state and industry disagreed over the amount of chemicals that leaked and their effects. Other companies admitted leaks only after environmental groups drew public and media attention to them.

One of these further leaks, that of phenol at a Ciba-Geigy plant, involved a runaway reaction.

This knock-on effect took place almost immediately as environmental groups went on the attack against the industry. Greenpeace charged Ciba-Geigy with having discharged high levels of the herbicide Atrazine into the Rhine for over a year. Ciba-Geigy responded by admitting a spill of Atrazine into the Rhine twenty-five hours before the Sandoz fire. Ciba-Geigy admitted the original Atrazine leak only after West German scientists had reported high Atrazine concentrations in the river. The government responded quickly to this charge. On 14 November the Swiss authorities confiscated Ciba-Geigy's production records for Atrazine from the Basle plant. This reflected a changed attitude on the part of the state: the previous day the Basle cantonal government had accused Sandoz of gross negligence. The director of the Basle water-protection agency, Benedict Hurni, said, 'We were never very sensitive about Atrazine before the Sandoz catastrophe. The Basle chemical industry should learn to take its responsibilities more seriously.' Ciba-Geigy eventually decided to cease production of Atrazine in Basle.

The string of chemical spills reported following the Sandoz fire also cast doubt on the chemical industry's reporting of spills previously. 'In the weeks following the Sandoz disaster more than fifteen accidental spills were reported: this is in sharp contrast to the fourteen reported accidents in the previous six years before the incident' (*Chemical Marketing Reporter*, 5 January 1987, p. 56).

West German Response

Chernobasle's knock-on effect was strongest in West Germany. Here again the question of reporting spills was raised, especially since some companies reported spills only after environmental groups drew attention to them. The two discharges by Bayer at the end of November were confirmed by the company only after environmental groups publicized them. The Social Democratic Party Environment Minister in the West German state of North Rhine-Westphalia said these accidents showed the need for strict punishment for firms withholding information on chemical accidents.

The West German Minister for the Environment responded: 'It is hard to believe that the fourteen accidents reported by the chemical industry under the regulation since its passage [in 1980] have been the only ones' (*European Chemical News*, 8 December 1986, p. 6). The Minister intended tightening the provisions on the reporting of chemical accidents. He was also contemplating ending self-regulation of industrial discharges and emissions.

In the immediate aftermath to the Sandoz spill, the West German chemical industry reacted by claiming 'West Germany possessed superior safeguards against a major chemical accident' (*Wall Street Journal Europe*, 24 November 1986, p. 4). The Verband der Chemischen Industrie, the West German chemical industry association, circulated in Bonn an internal report criticizing safety standards in the Swiss chemical industry generally. It also specifically accused Sandoz of violating existing legislation. The report said Sandoz housed 'water-sensitive material' such as phosgene in a building not designed for chemical storage. It also criticized the lack of drainage ducts, automatic sprinklers and heat and smoke alarms. The West German Social Democratic Party claimed on 9 November 1986 that safety measures at the Sandoz plant were inadequate, citing this report.

The industry position was supported after the Sandoz fire by the West German Environment Minister, who praised the West German chemical industry's unparalleled safety standards. The series of chemical spills and disputes that followed this unguarded expression of faith were to cause much embarrassment and to force the West German government to tighten rules on reporting chemical accidents. These spills caused much embarrassment to the chemical industry as well. The industry was in the middle of a major advertising campaign promoting the industry's success in environmental protection over the past decade. One advertisement began, 'Dear fish, you'll feel a lot better now that water pollution has been reduced'! The advertising campaign was hastily rejigged.

West German cartoon:
'No cause for alarm – as you see, it only happens in India.'

Disputes arose between state authorities and chemical company managements over certain spills. At the end of November, the state prosecutor's office investigated an unreported discharge of chlorobenzene into the River Main by Hoechst. The authorities suspected the spill was greater than the amount that Hoechst admitted was involved. In response to the leak of 2,4-D, BASF said it was only 'mildly toxic' and would dissolve within fourteen days in the Rhine's waters. Over the weekend after the spill BASF increased its estimate of the amount that had spilled from 2,420 lb to 6,400 lb, more than two tonnes. Independent experts said it could take twenty-six days or more to break down. Environmental officials said concentrations in the Rhine were 1,000 times higher than the 'safe level' and ordered waterworks closed. The officials were also reported to suspect that the spill contained even more toxic 'impurities'.

The political reaction in West Germany was very interesting. The Greens called for the 'complete detoxification of the chemical industry', involving a radical restructuring of the industry away from reliance on toxic products. They called for an end to the production of known cancer-causing chemicals, as well as the elimination of cadmium and the chlorinated hydrocarbons. The Social Democratic Party contented itself with calling for an investigation into the safety of all 90,000 chemical substances in current use. More radically, it proposed a major shift in the burden of proof regarding toxic injury. 'If a citizen has been damaged by the industry, it should be the industry, and not the victim, who should prove its guilt or innocence', said the Social Democratic Party spokesperson on the environment (*Wall Street Journal Europe*, 24 November 1986, p. 4). In the meantime, the Christian Democrat government announced it would back all individuals who claimed compensation for damage caused by the Sandoz spill.

The general industry fear was the introduction of yet more regulation of its activities. The director of the West German chemical industry association, Wolfgang Munde, said, 'We are against hasty passage of legislation which casts doubt on industrial society as a whole and against the drawing of parallels between Chernobyl, Bhopal and Sandoz' (*Wall Street Journal Europe*, 17 November 1986, p. 4). Indeed, industry spokespeople felt they had to defend the industry from the most basic attacks on its very existence. Sandoz board chairperson Marc Moret said: 'We must look at the total picture of chemistry and the environment, because a world without chemicals is unthinkable. It's part of our prosperity. We must organize it better and make it more secure' (*Wall Street Journal Europe*, 24 November 1986, p. 4).

The German chemical industry argued that none of the recent accidents could have been prevented by legal measures; it alleged: 'Today if someone poured a glass of milk into the Rhine it would be an environmental scandal.' Along with this toughening of response, the industry dug out a very old argument: 'Chemical companies have threatened to move production facilities abroad, thus eliminating badly needed jobs' (*Chemical Marketing Reporter*, 5 January 1987, pp. 3, 56). Along with this went the velvet glove of a reassurance campaign: the Verband der Chemischen Industrie argued: 'Sandoz was like an airplane disaster. It caused severe damage to the ecosystem. The spills at BASF, Hoechst and others are comparable to a car accident, with only minor damages to the metal' (*Chemical Marketing Reporter*, 5 January 1987, p. 57).

The Greens responded that the problem is related to the continued operation of the car; they argued that normal air and water discharges and emissions by the companies are the problem. As one example, the Greens claimed that the Bayer plant at Leverkusen can release sixteen tonnes of organic compounds into water every two hours. In the same period, the company can legally discharge 3.5 tonnes of ammonia compounds, 145 tonnes of sulphate, 550 g of mercury and 750 g of cadmium. The industry provides reports on these normal discharges to state bodies, but they are not available to the public. Pressure is now being exerted to force release of the actual records of emissions and discharges. Thus, as in the USA, the knock-on effect of Chernobasle drew public attention and disquiet not only to accidental but also to normal discharges and emissions by the chemical industry.

Given the continued development of Green politics in West Germany, and the attempt by the establishment parties to drape themselves also in green, increased regulation of the chemical industry there is inevitable. On a more general level, the industry fears increased regulation by the EEC. The industry was reported to be extremely suspicious at 1987 being named European Year of the Environment by the EEC. Industry spokespeople were worried the EEC Commission might use this as an attempt to frighten the general public into accepting a new round of regulation. The industry position is that the Commission should wait to see the effect of the Seveso directive before embarking on a new round of regulation. This could take some time, given that the Commission is satisfied with only three countries' legislation to implement the directive: Denmark, France, and Britain.

10 Against Toxic Capital

'We in the business world have learned that we must consider the environmental impact of economic goals. The time is here for environmentalists to consider the economic impact of environmental goals.' – Jackson Browning, UCC

'Ideally I'd like to see this whole joint closed down. The company sucks. It's all unsafe – the fumes and all that.' – Chemical worker, Elizabeth, New Jersey (quoted in Halle, 1984, p. 114)

'Can you think of a better solution than to get out of the chemical industry?' – Rudolf Bahro

The management of the Bhopal crisis by capital and state is an indication of the strengths and weaknesses internationally of anti-toxic opposition. While the industry prevented a major setback, it did not achieve a total victory. The resurfacing of its safety crisis in the USA and in Europe indicates that this one will run and run. Indeed, it can be argued that toxic capital has continuously faced severe health and safety crises for the past two decades. The deepening of this crisis, shown by Bhopal, Ixhuatapec and Cubatão, was confirmed by the near-Bhopals of Institute and Chernobasle.

Opposition to toxic capital has until now been hampered by the local nature of its struggles. This is inevitable, given that vigorous opposition occurs mainly in two circumstances: either an obvious insult has been suffered and connected with its toxic source, as at Love Canal; or a new toxic plan has been announced for an area – as at Phuket, Livingston and Warren County.[1] Thus anti-toxic response to Bhopal was localized and varied. Although struggle was of course most intense in Bhopal itself, reaction was strong throughout Asia. In Europe, only after Chernobasle did the threat of toxic chemicals sink home. In the USA a strong response came both immediately after Bhopal and since then, because of the continuing high level of struggle against toxic capital by local community groups and national environmental action groups. Bhopal continues to find echoes and provides a rallying cry in many struggles around the world.

For anti-toxic opposition to widen, we must make clear the connections between all the different struggles against toxic capital. We need to emphasize various themes that have previously appeared at various stages and localities of anti-toxic struggle but which all were to reappear in the Bhopal crisis. Following that, we need to look at the general safety crisis of toxic capital, at how we can contribute to widening and deepening that crisis, in order to break the power of toxic capital to poison us.

Learning from Bhopal

The centrality of women in toxic struggles was confirmed by Bhopal. Women played a major role in the agitation in Bhopal itself and were prominent members of the various opposition groups. This confirms what activists already know from their own campaigning: women are perhaps the group most ready to attack toxic capital. In some ways, this reaction is one of self-defence, given the vicious attacks toxic capital has launched on women. Here we must consider not only the genetic and reproductive insults women have suffered, but also the creation of additional unpaid labour for women, including emotional labour, and the personal toll these stresses extract from women. In Bhopal the state attempted to deal with the gas leak's reproductive effects by limiting them to possible teratogenicity – the production of deformed or less able children. This attempt at limitation did not go unchallenged: the immediate reproductive effects on women's health were major and were investigated and highlighted by opposition groups.

One cautionary note must be sounded here. There has been a tendency to see reproductive effects of toxic chemicals as purely a women's issue.

This view is misleading. Not only do men share in the responsibility for children – but also, toxic chemicals that affect women's reproductive health do not bypass the male reproductive system. In Bhopal, for example, impotence and loss of libido were reported among a large proportion of exposed males. Reproductive effects are a transgender issue that men as well as women should address.

Bhopal also highlighted the lack of available information, even on extremely hazardous chemicals. It also demonstrated the deleterious effects of property rights on information about public health. This was again highlighted by the lack of information on aldicarb oxime, the chemical involved in the Institute leak, and the slowness of Sandoz in providing information on what chemicals were involved in Chernobasle. The total failure of UCC to warn workers, local community and the Indian government is just one more episode from the industry's long history of censorship, suppression of information and dissemination of misinformation. The industry's policy of classifying toxicological information as trade secrets was again exposed as a threat to public health and emphasized once again the need for free access to all information in this area.

The closeness of chemicals for peace to chemicals for war was shown in the constant comparisons of post-killing Bhopal with war zones, particularly with Hiroshima and Nagasaki.[2] This closeness was also emphasized by the allegations regarding chemical warfare at UCIL's research and development centre in Bhopal, Dr Ballantyne's ties with the chemical-warfare establishment and the similarity of action between MIC and chemical-warfare agents.[3] Also, of course, the pesticides produced at Bhopal are themselves part of a war, the war against the land involved in agribusiness, which Engler has rightly characterized as an extractive industry, with all that that implies.

Bhopal has also confirmed some political lessons. While reformists have generally argued for increased state regulation of toxic industry, radicals have argued that such government regulation is ineffective and provides only reassurances of safety, rather than the reality of safety. At Bhopal, the Indian state failed completely in its duty to regulate a hazardous plant. This failure was due not only to the corruption of the local state, but also to the shared interest of both state and capital in the project of 'development'.

Again, at Institute and Chernobasle, the state remained unable to protect the local population.

No One is Safe

This means that serious toxic hazards exist throughout the world, no matter what the level of state regulation. While Castleman and Purkayastha have adequately documented the double standards in operation at Bhopal and Institute, the problem is not reducible to hazard export from the West, for at least two reasons: first, there is a racist distribution of risks within the metropolitan countries; and second, even the conditions at the Bhopal plant would have led to no regulatory action in the USA, the most highly regulated country in the world.

If our analysis stops at Bhopal, not only does Bhopal become isolated but we also miss out on the connections between Bhopal and the normal operations of toxic capital. We should see the racist character of the Bhopal plant as a special case of the generally oppressive design and operation of chemical-process industries as a whole. We need to emphasize the politics of this by linking Bhopal to Institute:

While the double standard does exist in some cases, the more fundamental issue of safety of hazardous chemical facilities world-wide seems more basic. A recent report by the International Confederation of Free Trade Unions argues that despite the technical design flaws and poor safety and maintenance records at the Bhopal facility, adherence to US standards would not have averted this disaster. The accident didn't happen because we're safe and they're not. No one is safe. (Wartenburg, 1985)

This would appear to be a more fruitful argument politically than the hazard export one (Jones, 1986). In responding to Bhopal, it is essential that we do not treat it in isolation. This book has concentrated almost exclusively on only one danger posed by toxic chemicals: the danger of accidents in major chemical-processing plants or storage areas releasing clouds of toxic chemicals. Yet even with this limitation of scope, it was impossible to avoid the knock-on effect these accidents had in metropolitan countries on normal toxic-chemical emissions. It is necessary to extend the critique of toxic-chemical use to all parts of the toxic cycle. In this way

Bhopal can be seen as the extreme end of a continuum which stretches to smaller accidents (only three dead, put it on page 27), to the routine release of carcinogenic chemicals into the environment, to the slow murder of hundreds of thousands of workers through 'low-level' exposure to chemicals at work, to the contamination and toxic nature of consumer products.

This perspective implies that we must be careful not to allow concern with possible disasters to deflect attention away from other areas of insult by toxic capital. Here we may take anti-nuclear opposition as an example. While that opposition has made use of the threat of a catastrophic release of radiation through a melt-down, its major focus has been on the continual release of 'low-level' radiation. We need to make the public aware that the cancer rings around chemical plants are as lethal as the cancer rings around nuclear-power stations.

We need to emphasize both the chronic and acute aspects of toxic chemicals. As Sujatha Gothoskar of the Union Research Group says, 'But what we are finding out and exposing about Union Carbide today, should also be done about other hazardous plants and industries. We cannot afford to awaken to disasters alone.' As Russell Mokhiber, lawyer with the Corporate Accountability Research Group observed: 'The Bhopal disaster is exceptional, not because corporate negligence and recklessness are unusual at Union Carbide but because of the magnitude of the suffering in this case.'

It is also important that we do not focus exclusively on multinational capital. An equal or even greater danger is posed by indigenous, and particularly state-owned, capital: Pemex in Mexico, Petrobras in Brazil, Nitrigin Eireann Teoranta in Ireland, Nypro in Britain[4] and various public-sector companies in India. What we must attack is the material content of toxic production, rather than the type of capital involved.

We must also stress the inevitability of accidents: we must emphasize that accidents are built into the technology itself – in its widest sense, which includes the social relations built into the technology – and thus cannot be resolved by technical fixes, however elaborate and costly.

Exterminism

Perhaps the most disgusting response to Bhopal has been capital's attempt to write off the thousands of corpses as the cost of increased food production which has enabled India's 'teeming millions' to survive. This is the cost–benefit argument at its most blatant. The *Wall Street Journal* editorialized: 'Of the people killed, half would not have been alive were it not for that plant and the modern health standards made possible by the use of pesticides.' The *Guardian* rowed in with rhetoric on the balance-sheet of death and the *New York Times* spoke of the pain of progress. Thus Bhopal was presented as the cost that must be borne for the benefit of increased food production. Yet the benefit in this case is mythical. The mythical view being endorsed in those newspapers is that increased pesticide use, as part of the application of Western agricultural technology in what has been misleadingly called the Green Revolution, has increased the availability of food for the people of the peripheral countries. This myth has been adequately debunked elsewhere (George, S., 1976; and Everest, 1985, chapter 5). That is, the integration of peripheral countries' agriculture into the international business system, with the resulting changes in land ownership and use, destroyed subsistence farming in those countries and created the major problem of access to land and food. The newspapers are regurgitating what is essentially propaganda for a fertilizer/pesticide sales project.

Bhopal shows blatantly the nature of development. As one Indian observer noted: 'The benefits of twenty-first century technology will go to politicians, scientists and so-called intellectuals. But the people who have to pay the price of this technology with their lives are those who cannot even get enough to eat two times a day.' As R. Mayur, professor of geography at the University of Reading in England, observed, 'Most of the people who died that night were hardly beneficiaries of the industrialization proceeding near their shanties.'

We need to widen this critique of development to include industrial development in the metropolitan countries. As Engler argues (1985, p. 489): 'A complete history of American mining, foundries, mills and farming, along with a host of other industries, would yield a staggering casualty list. Yet this continuing chemical warfare against the American people has remained largely unanalysed; it is considered an incidental cost of growth or played down as hysteria.' This death by toxic development takes a slower and more subtle form in the metropolises, though the potential for disasters also exists there. When asked if a Bhopal-type disaster could occur in the USA, Samuel Epstein replied; 'I think what we're seeing in America is far more slow – not such large accidental occurrences, but a slow, gradual leakage with the result you have excess cancers or reproductive abnormalities.'

These comments link in with the general ecological critique of capitalist production. The West German Green, Rudolf Bahro, among others,[5] has characterized the present stage of industrial development as a form of exterminism. This concept of exterminism comes from E. P. Thompson, the British historian and activist against nuclear weapons.[6] Thompson coined the term to refer to 'the tendency to self-extinction of the human race that is present in the growing independence of the arms race from even the rational interests of its most powerful protagonists'. Bahro has extended the term to the industrial system in general: 'Inseparably connected to the military and economic aggression, exterminism is expressed in the destruction of the natural basis of our existence as a species'(1982, p. 124).[7]

I share Bahro's position. The present state of industrial development is ecocidal and therefore genocidal. Just as nuclear weapons present the possibility of ending the existence of human beings, so the widespread destruction, contamination and manipulation of the biosphere by toxic capital presents a major and growing threat to the continuation of human life on earth. Rather than attempt a comprehensive survey, here I will examine one particular aspect of this exterminism, a widespread slow-motion Bhopal which is often ignored by ecologists – work-place exposure to toxic chemicals. The review will be brief and confined to the USA.

Corporate Killing in the Work-place

About 21 million Americans – or one out of every five workers – are exposed to hazardous substances on the job, and more than twice that many are exposed to such substances some time during their working life. One out of four Americans will suffer from cancer during his or her lifetime, and up to 38 percent of all cancers in the USA are related to substances in the work-

place. Between 8 and 11 million workers have been exposed to one cancer-causing substance, asbestos, since the Second World War, with more than 2 million of those expected to die from asbestos-related cancer. Two million workers face exposure to benzene, a highly toxic chemical which can cause leukaemia. Another 1.5 million workers are exposed to arsenic, also a cancer threat. Three out of four miners now receiving a pension have an irreversible lung disease caused by coal dust. About 800,000 cases of job-related skin disorders occur each year as the result of exposure to toxic substances. (Witt, 1979, p. 165)[8]

Major controversy has developed in the USA and elsewhere over how great the responsibility of work-place exposure to toxic substances is for the incidence of cancer. In looking at possible chemical causes or factors in such diseases, one major difficulty is that chemical exposure is now so widespread and so mixed that it is hard to isolate particular chemicals and connect them with deleterious health effects. The chemicals and toxic substances that have been proven to be hazardous have normally been ones that produce specific, large increases in rare cancers and diseases or have identifiable diseases specific to them, for example, vinyl chloride and angiosarcoma and asbestos and asbestosis and mesothelioma. While many chemicals have been identified as causes of cancer in laboratory tests on animals, surveys of their effects in human populations (epidemiological surveys) are scarce: 'A large number of industrial materials have been identified as carcinogens in laboratory studies, but have not been investigated in human populations . . . most animal carcinogens now in commerce have not been and are not likely to be subject to epidemiological study. For these substances, waiting for epidemiological confirmation becomes waiting for Godot' (Davis, D. L., *et al.*, 1981, p. 289).

Yet capital constantly argues that animal evidence is not enough to convict a chemical. Toxic chemicals should be given the benefit of the doubt, they should not be abandoned or regulated until there is proof of their ill-effects in humans. This has been coarsely described as a 'show me the bodies' attitude. Capital has argued for scientifically reliable studies showing health effects on humans before they will accept a chemical is hazardous. This argument is convenient, as it takes some twenty to thirty years before the cancer-causing effect of a chemical becomes evident in people. By that stage the all-important highly profitable monopoly stage of production has passed.

We have already argued that a major increase in production of toxic chemicals has taken place over the last few decades. We are therefore likely to see major increases in cancers and other diseases caused by this increased production. 'Production and normal use of many carcinogenic, bioavailable[9] and otherwise hazardous substances has reached very high levels only since the early 1960s' (Davis, D. L., *et al.*, 1981, p. 293). What will be the effect of those chemicals about which we know little or nothing? It can only be a subject for speculation: 'Fewer than 10 percent of all chemicals in commerce have been even minimally tested' (Davis D. L., *et al.*, 1981, p. 293). We should also recall here that, to minimize ill-effects, many such tests have been shown to be fraudulent or 'massaged' by the testing companies and the chemical companies themselves.

As to the overall effect, Joseph Wagoner, of the US National Institute for Occupational Safety and Health, warned in 1976: 'It seems clear that society may soon be faced with a public-health hazard of monumental and perhaps irreversible proportions' (Wagoner, p. 3). Irving Selikoff, who was responsible for the exposure of the health effects of asbestos in the early 1970s, has pointed out that if only one of the chemicals introduced in the 1960s proves to be as toxic as asbestos, the current epidemic of occupational cancer will continue for decades into the future. As D. L. Davis *et al.* (p. 308) point out, 'If as few as 5 percent of all new cancers in the USA are related to industrial exposure, this would still represent a major public-health problem resulting in 20,000 excess cancer deaths each year.'

It's important to emphasize that this toxic exposure at work is not confined to the chemical industry proper. Appendix 1 ('Some Toxic Hazards') at the end of this chapter indicates the wide spread of carcinogenic exposure throughout industry; as the list dates from the early 1980s, other examples could certainly be added. Perhaps the best way of illustrating the widespread nature of these threats at work is to glance at the situation in the electronics industry. Electronics has created for itself the image of an ultra-clean industry. This image began to fall apart in 1982 when pollution in Silicon Valley was linked to increased rates of reproductive problems among residents. Since then problems related to pollution, toxic-waste disposal and work-place hazards in the electronics industry have been increasingly reported. The *New York Times* reported in 1984 (10 November, p. 44): '... statistics suggest that the incidence of work-related illnesses among semiconductor workers is triple the average for all of [California's] manufacturing workers. And over the last four years, the illness rate for semiconductor workers has consistently exceeded the rate for workers in the mining, construction, metals or chemicals industries.'

In January 1987 a study commissioned by Digital Equipment Corporation was reported to have found women workers in the etching and gas treatment sections of Digital's semiconductors' lines had a miscarriage rate of 39 percent, compared with the US national average of 20 percent. This study confirmed earlier anecdotal accounts of increased reproduction problems for women workers in the electronics industry. It also exemplifies another major toxic hazard – the effect of toxic substances on reproduction. Effects include menstrual irregularity, increased incidence of abortion, congenital malformation, foetal-growth retardation, foetal and postnatal carcinogenesis, decreased libido and sterility.[10] Tony Mazzocchi, ex-safety officer of the Oil, Chemical and Atomic Workers' Union, has indicated what is involved: 'Workers are going to find out that because of their labour the children they worked for and hoped would have a better life, will instead have abnormalities. There never was an era like the eighties will be. People will see the damage that has been done' (quoted in Jackson, 1986, p. 21).

This general toxic crisis is a direct result of capitalist industrial development. As Joseph Eyer has noted: 'Economic growth – capitalism – is pathogenic.' This development, in Bahro's words, 'is driven by a boundless need to valorize capital, to make value into more value.' The major problems lie in the inherent structure and expansionist urge of industry. This is most obvious in the chemical industry. The nature of the industry ties profitability to the continuous introduction of new products. The first five years or so of the introduction of a new chemical product allows the firm introducing that product to make extremely high profits from its effective monopoly position. As early as 1979, Barry Commoner tied the major problem of pollution to the economic structure of the chemical industry:

Thus, the extraordinarily high rate of profit of this industry appears to be a direct result of the development and production at rapid intervals of new, usually unnatural, synthetic materials – which, entering the environment, for reasons already given, often pollute it. This situation is an ecologist's nightmare, for in the four-to-five-year period in which a new synthetic substance, such as a detergent or pesticide, is massively moved into the market – and into the environment – there is literally not enough time to work out its ecological effects. Inevitably, by the time the effects are known the damage is done and the inertia of the heavy investment in a new productive technology makes a retreat extraordinarily difficult. The very system of enhancing profit in this industry is precisely the cause of its intense, detrimental impact on the environment. (1979, p. 260)

Exterminism is the cul-de-sac down which capital is driving us. Like the production of nuclear weapons, the use of toxic chemicals is beyond rational control. The runaway reactions that have featured in this book are literally toxic capital out of control. Rationally, capital should be concerned with the long-term effects of chemical contamination. In the same way, it should be concerned with the present lunatic schemes for tropical forest development which will wipe out the major remaining source of new genetic capital. Yet capital is amazingly blind to its rational, long-term interests and increasingly focused on the 'quick buck'. The irrationality of capital's present plans has been previously pointed out in the massive gamble it contemplated in nuclear power production.[11] A similar irrationality pervades the continued introduction and use of highly toxic chemicals. As some anti-toxic campaigners have argued, 'If the nuclear state is the end of the logic of growth, then the toxic state is the growth of the end of logic' (*Rebel*, 14, 1979, p. 2).

Attacking Toxic Capital

The International Confederation of Free Trade Unions produced an excellent international trade-union report which concluded that the conditions that led to the killing at Bhopal exist at chemical plants throughout the world; yet the report calls only for increased regulation, the passing of recommendations at the International Labour Organization and other international organizations, increased communications between trade unions world-wide and the like. Trade unions – like the Oil, Chemical and Atomic Workers' Union in Geismar in the USA and the General, Municipal, Boilermakers' and Allied Trades Union in Britain – attack the chemical industry on safety and ally themselves with ecological and community groups; but their perspectives still remain limited to increased staffing, regulation and safety measures. While these measures are welcome, they will not resolve the basic problem.

The response from the left, supposedly unaffected by the limitations of a trade-union consciousness, has been as poor. It is indicative of the poverty of the left, as well as its workerist fixation, that its response mainly centred on support for the struggle of the UCIL workers in Bhopal. Tied to the ideology of progress through industrial and technological development, and given its view of itself as an alternative manager of this development, the left has little desire to examine radically the implications of Bhopal, despite its recent fascination with all things green since the recent electoral successes of some Green parties in Western Europe. It's worth quoting Bahro at some length here on the left's general failure:

Even that supposedly progressive economic analysis which uses Marxist categories functions today in conformity with the system. It goes on feeling the pulse of a still continuing accumulation of capital, calculates profit rates and forecasts short-term – and recently also long-term – crises. But it has nothing more to say on how this pulse is to be stopped, how the accumulation of capital can be not just measured but actually brought to an end. All that is left is the latest economistic reformism, which already assumes the next long wave, the breakthrough into eco- and bio-industries, total cable communication, etc., as an inadvertible given which we have to surrender and adapt ourselves to. (Bahro, 1982, p. 146)

The absence of the left should not be a major cause for dismay, however. Bahro has pointed out that the politically most conservative forces among the West German Greens have been the various leftist groups that have joined them.

In practical terms, it appears that future anti-toxic tactics will focus on demanding freedom of information and increasing capital's costs. While information is a possible major source of profit and control for capital, it can also be a major weapon in blocking toxic capital's plans. Depending on the situation, demands will vary from the basic one of freedom of information to criticism of the type and level of information toxic capital provides. In EEC countries, the information provisions of the Seveso directive should be used, while their extension to all chemical plants should be demanded. (The British 'Control of Major Industrial Hazards' regulations, for example, apply only to about 250 of the 1,500 hazardous sites notified to the Health and Safety Executive.)

It's important to emphasize the aspect that the industry most fears: reporting requirements on ordinary non-accidental emissions. Since capital is afraid here, it should be attacked on this flank. The tactics will be to increase the industry's costs across the board, not only by demanding increased regulation and provision of information, but also by contesting the activities of toxic capital locally.

It must be emphasized that reformist demands for increased regulation, while sometimes useful in driving up industry's costs, in no way provide an adequate response to the toxic crisis. For example, consider Wolfe's analysis of twenty-two regulatory actions on carcinogens in the USA, reportedly the most heavily regulated industrial country. He showed that only four actions were begun by the government regulatory apparatus: the others were forced on the regulators by trade unions and environmental and consumer organisations (Wolfe, 1977). And his analysis was undertaken before the recent dismantling of regulation under the first Reagan regime.

Moreover, regulation can be useful for toxic capital: while restraining the most blatant abuses, it reassures the public that the industry is being controlled and everything's OK. In that spirit, some sections of US toxic capital called for more effective and obvious regulation towards the end of the first Reagan regime. Regulation must not only be done, it must be seen to be done.

Our strategy in attacking toxic capital must be to increase its costs, delay its projects and block its products – while at the same time arguing against the ideology of development and progress, by underlining the dangers inherent in continued toxic development. Regarding chemical plants specifically, it is necessary to universalize struggles against their operation by mounting a general critique of the industry, rather than going from case to case. This general critique will be in some senses similar to the general critique of nuclear power, which is focused not on one plant, but on the whole industry. Our project should be the same as the anti-nuclear project: to stop production.

Of course, the difficulty in anti-toxic work is that we are dealing with a plethora of chemicals and other substances, with varying dangers and exposure routes, which is not like anti-nuclear work, where the whole debate is focused on radiation. Chemicals are also more basic to capital than nuclear power, which, however essential it is to the military machine, is in social terms just one of several energy sources. In comparison, toxic chemicals are basic to the continued existence of capital's industrial production and thus to the continued accumulation of capital. Thus a critique of toxic production questions the whole base of capital accumulation.

It is necessary to deepen and extend this critique and, with it, anti-toxic opposition. In doing this 'we must question the process by which poisonous chemicals, like MIC, have become intrinsic parts of our lives and whether we need them. We must question the economic system, more and more an internationalized one, which creates the chemicals, dominates their markets and influences local development efforts' (Agarwal, 1985, p. 29).

Work against toxic capital must necessarily operate on two levels. On the practical level, it is essential to attack and, if possible, block or impede the specific activities of individual centres of toxic capital. These attacks can focus on all parts of the toxic cycle – at the point of production, on disposal of waste products, on the toxicity of consumer products and on transportation of toxins. Given the mass murder that toxic capital entails, our attacks on it are the only logical form of self-defence. On another level, anti-toxic work requires a critique of the material content of capital's present production. This critique must be the basis of attacks on the ideology and propaganda of toxic capital. Toxic capital has argued, in Monsanto's famous advertising slogan, 'Without chemicals, life itself is impossible' – by which they mean that capitalism is impossible. Yet, as we have seen, the chemicals necessary for capitalism are making life itself impossible.

The choice is becoming starker year by year, disaster by disaster. Toxic capital stands in global and antagonistic contradiction to the continuation of life on this planet. Victory over toxic capital holds major liberatory possibilities: as one American comrade who was sprayed with Agent

Orange in Vietnam observed at the last Carnsore Point anti-nuclear festival in Ireland: 'Without petrochemicals, capitalism itself is impossible.'

APPENDIX 1
SOME TOXIC HAZARDS

The following is a run-through of only some of the general hazards associated with various materials and sectors of toxic industry:

MATERIALS

Leather and leather products. Increased risk of bladder and mouth cancer in males; increase in blood cancers (lymphomas) in males and females. Cancer of nose and sinuses in boot and shoe industry. Exposure to benzene.
Petrochemicals. Increase in stomach and brain cancer. Threefold increase in oesophageal and stomach cancer and double risk of lung cancer in oil-refinery workers. Exposure to benexene. Doubled risk of cancer to the intestines and other digestive organs (refinery workers). Nasal and skin cancer.
Plastics. Exposure to the following plastics is suspected of causing cancer: Cellophane, dextran, polyacrylics, polyamides, polyethylene, polystyrene, polyurethanes, polyvinyl, pymolidone. The following plastics are known to cause cancer: acrylonitrile, PVC. There is also risk from exposure to many untested and suspect chemical substances such as antioxidants,

flame retardants, plasticizers and stabilizers. Genetic risks are also associated with some of these.

Rubber. Increased risk of stomach and prostate cancer. In rubber- and cable-making, high incidence of bladder, lung and stomach cancer. In tyre sector, high risk of stomach and bladder cancer; also leukaemia. Bronchitis caused by tyre-curing fumes. Synthetic rubber: increased risk of leukaemia suspected from substances involved (benzene, butadiene, styrene).

Textiles. High incidence of lung cancer. Also other respiratory diseases.

Industrial Sector

Electronics. Asthma. Also: exposure to solvents.

Furniture-making. Lacquerers face breathing and nervous problems. Exposure to solvents.

Iron and steel-making. Increased risk of lung cancer for furnacemen, rollers and skilled assistants: metal moulders, casters, iron-foundry furnacemen and labourers.

Metal-plating and polishing. Increased risk of oesophageal cancer and primary liver cancer. Also cancer of nose, lung and larynx. Electroplating: hazards from hydrocyanic acid, cyanides and cyanic compounds, mercury and its compounds, ammonia, nitric acid, solvents and nickel and its compounds.

Painting and decorating. Exposure to pigments, solvents and other chemicals in paints. Excess risk of respiratory and gall-bladder cancer.

Printing. Increased risk of death from cancer of the lung, large bowel, oral cavity, bladder and blood (leukaemia). The English Print Union has produced a list of eighty-three chemicals, suspected of causing cancer, used in the industry. Dermatitis risk from some inks.

Shipyards. Allergies (to epoxies) are common; there is often severe hearing loss. Increased respiratory disease among welders.

Smelting. Copper, lead and zinc: increased cancer risk owing to arsenic exposure. Aluminium: lung cancer and terminal leukaemia.

Services Sector

Drycleaners. Dermatitis, liver, kidney and brain damage. Cancer risk. Exposure to variety of solvents, especially perchloroethylene (Perklone).

Hairdressers. Increased risk of lung cancer, also other respiratory diseases. Exposure to hair dyes, aerosols, talc, heat, ultraviolet light.

Hospital and dental workers. Risk of infection, exposure to radiation and toxic chemicals. Danger to pregnant operating-room workers from exposure to anaesthetic gases. Also exposure to disinfectants and sterilizing agents.

Laboratory workers. Women workers in chemical laboratories face higher risk of miscarriage, prenatal death and major malformations in their children. Also increased risk of cancers, particularly of the blood (leukaemia and lymphomas).

Office workers. Exposure to solvents, visual display units (VDUs), fumes from photocopiers, microfiches. Many eye problems.

APPENDIX 2
MAJOR HAZARD SITES IN GREAT BRITAIN

In 1986 the following list was released by the Health and Safety Executive and published in the May issue of *Health & Safety at Work*. It included full postal addresses. It is reprinted by kind permission of the Health and Safety Executive. The list has not been updated since 1986, so it is possible that certain establishments should be added or dropped.

East Anglia

British Gas Corporation, Canvey Island, Essex. **British Gas Corporation,** Bacton, Norfolk. **Calor Gas Ltd,** Stanford-le-Hope, Essex. **Calor Gas Ltd,** Bury St Edmunds, Suffolk. **Calor Gas Ltd,** Felixstowe, Suffolk. **Carless Solvents Ltd,** Harwich, Essex. **Dow Chemicals Ltd,** King's Lynn, Norfolk. **International Marine Management (Bond) Ltd,** Felixstowe, Suffolk. **Leisureking Ltd,** Bury St Edmunds, Suffolk. **London and Coastal Oil Wharves Ltd,** Canvey Island, Essex. **Mobil Oil Co. Ltd,** Coryton, Essex. **Obel Warehousing Ltd,** Bury St Edmunds, Suffolk. **Phillips Petroleum Co. Ltd,** Bacton, Norfolk. **Powell Duffryn Terminals Ltd,** Purfleet, Essex. **Shell UK Oil Refining & Marketing Co. Ltd,** Stanford-

le-Hope, Essex. **Unitank Storage Co. Ltd**, West Thurrock, Grays, Essex.
E. G. Wright Ltd, Saffron Walden, Essex.

Northern Home Counties

Anglian Water Authority, Huntingdon, Cambridgeshire. **Firmin Coates Ltd**, Aylesbury, Bucks.

London

ACC Gases Ltd, Enfield, London. **Berk Spencer Acids Ltd**, Stratford, London. **British Gas Corporation**, Fulham, London. **British Gas Corporation**, Southall Holder Station, Middlesex. **British Gas (North Thames)**, Bromley-by-Bow, London. **May and Baker Belvedere Ltd**, Belvedere, Kent. **North Thames Gas**, Beckton, London.

Manchester

British Gas (NW), Miles Platting, Manchester. **British Gas (NW)**, Partington, Manchester. **Courtaulds Ltd**, Trafford Park, Manchester. **Robert Fletcher & Sons Ltd**, Greenfield, Oldham. **Shell Chemicals UK Ltd**, Urmston, Manchester.

Merseyside and Cheshire

Associated Octel Co. Ltd, South Wirral, Cheshire. **Calor Gas Ltd**, South Wirral, Cheshire. **Chlorchem Ltd**, Widnes, Cheshire. **Firmin Coates & Sons Ltd**, Middlewich, Cheshire. **Hays Chemicals Ltd**, Sandbach, Cheshire. **Imperial Chemical Industries plc**, Northwich, Cheshire. **Imperial Chemical Industries plc**, Runcorn, Cheshire. **Imperial Chemical Industries plc**, Widnes, Cheshire. **Marchem Ltd**, Widnes, Cheshire. **North West Gas**, Northwich, Cheshire. **Petrolite Ltd**, Kirkby, Liverpool. **Shell Chemicals (UK) Ltd**, South Wirral, Cheshire. **Shell UK Oil**, Knutsford, Cheshire. **Shell UK Oil**, South Wirral, Cheshire. **Shell UK Oil**, Birkenhead, Merseyside. **UKF Fertilisers Ltd**, Chester, Cheshire. **Unitank Storage Co. Ltd**, Eastham, Merseyside. **Ward Blenkinsop & Co. Ltd**, Widnes, Cheshire.

East Midlands

Calor Gas Ltd, Stoney Stanton, Leics. **Camping Gaz (GB) Ltd**, Sywell, Northants. **Coalite Fuels and Chemicals Ltd**, Bolsover, Derbyshire. **Conoco Ltd**, Grimsby, Lincs. **Courtaulds Acetate Ltd**, Spondon, Derby. **Everard & Saltmarsh Ltd**, Stamford, Lincs. **Flogas Ltd**, Merrylees, Leicester. **Freedom LPG Ltd**, Staveley, Derbyshire. **Knights of Old Ltd**, Cherry Hill, Northants. **Newark Storage Co. Ltd**, Newark, Notts. **Norsk Hydro Polymers Ltd**, Staveley, Derbyshire. **Reckitt Household and Toiletry Products Ltd**, Sinfin Lane, Derby. **Staveley Chemicals Ltd**, Staveley, Derbyshire. **Super Gas Ltd**, Witney, Oxford. **W. H. Wallington & Sons**, Chipping Norton, Oxford.

West Midlands

Albright & Wilson Ltd, Warley, West Midlands. **British Gas (W. Mid.)**, Aston Church Road, Birmingham. **British Gas (W. Mid.)**, Windsor Street, Birmingham. **British Gas (W. Mid.)**, Foleshill Road, Coventry. **Blockleys Ltd**, Telford, Salop. **Esso Petroleum Co. Ltd**, Tunstall, Stoke-on-Trent. **Hereford Storage Ltd**, Hereford. **Hereford Storage Ltd**, Shrewsbury. **Hereford Storage Ltd**, Lichfield, Staffs. **Hereford Storage Ltd**, Telford, Salop. **A. J. Maiden**, Telford, Salop. **Shell UK Oil**, Warley, West Midlands.

North-east

BASF Chemicals Ltd, Middlesbrough, Cleveland. **British Gas (Northern)**, Middlesbrough, Cleveland. **Calor Gas Ltd**, Port Clarence, Cleveland. **R. Durham & Sons**, Billingham, Cleveland. **Durham Chemicals Ltd**, Birtley, Tyne and Wear. **Imperial Chemical Industries plc**, Billingham, Cleveland. **Imperial Chemical Industries plc**, (ICI Petrochemicals), Middlesbrough, Cleveland. **Imperial Chemical Industries plc**, Port Clarence, Cleveland. **Imperial Chemical Industries plc**, Wilton, Cleveland. **Norsk Hydro Polymers Ltd**, Newton Aycliff, Durham. **Phillips Petroleum Co. (UK) Ltd**, Hartlepool, Cleveland. **Phillips Petroleum Co. Ltd**, Middlesbrough, Cleveland. **Seal Sands Storage Ltd**, Middlesbrough, Cleveland. **Tees Storage Co. Ltd**, Middlesbrough,

Cleveland. **Thermogas Ltd**, Blaydon, Newcastle. **Tioxide UK Ltd**, Hartlepool, Cleveland.

NORTH-WEST

BCL Ltd, Barrow-in-Furness, Cumbria. **British Gas Corporation**, Barrow-in-Furness, Cumbria. **Carlisle Warehousing Ltd**, Carlisle, Cumbria. **Imperial Chemical Industries plc**, Thornton, Cleveleys, Lancs. **Imperial Chemical Industries plc**, Morecambe, Lancs.

SOUTH-EAST

BP Oil Kent Refinery Ltd, Rochester, Kent. **British Gas Corporation**, Rochester, Kent. **Johnson Wax Ltd**, Camberley, Surrey. **Redland Bricks Ltd**, Horsham, West Sussex.

SOUTH-WEST

Aerosols International Ltd, Wellington, Somerset. **British Cellophane Ltd**, Bridgwater, Somerset. **British Gas LNG Facilities**, Hallen, Bristol. **Calor Gas Ltd**, Ivybridge, Devon. **Calor Gas Ltd**, Plymouth, Devon. **Imperial Chemical Industries plc**, (ICI Agricultural Division), Avonmouth, Bristol. **ISC Chemicals Ltd**, Avonmouth, Bristol. **Norsk Hydro Fertilisers Ltd**, Avonmouth, Bristol. **Tenneco Organics Ltd**, Avonmouth, Bristol.

SOUTHAMPTON AREA

Calor Gas Ltd, Southampton, Hants. **EniChem Elastomers Ltd**, Hythe, Southampton. **Esso Chemicals Ltd**, Hythe, Southampton. **Esso Petroleum Ltd**, (Esso Refinery), Fawley, Southampton. **Hythe Chemicals Ltd**, Hythe, Southampton.

SOUTH YORKSHIRE AND HUMBERSIDE

BP Chemicals Ltd, Hull, Humberside. **Britag (Agricultural Sector)**, Burton-on-Humber, Humberside. **Britannia Gas – LPG Terminal**, Immingham, Humberside. **British Gas Corporation**, Driffield, Humberside. **British Gas Corporation**, Easington, Humberside. **British Gas Corporation**, Grimsby, Humberside. **British Gas (East Midlands)**, Barrow Road, Sheffield. **Calor Gas Ltd**, Immingham, Humberside. **Conoco Ltd**, Immingham, Humberside. **Conoco Marketing Ltd**, Immingham, Humberside. **Courtaulds plc**, Grimsby, Humberside. **Doverstrand Ltd**, Immingham, Humberside. **Euroshires Ltd**, Brigg, Humberside. **Hepworth Pipe Co. Ltd**, (Hepworth Iron & Steel), Stockbridge, Sheffield. **Immingham Storage Ltd**, Immingham, Humberside. **Linsey Oil Refinery Ltd**, Immingham, Humberside. **Norsk Hydro Fertilisers Ltd**, Immingham, Humberside. **SCM Chemicals**, Immingham, Humberside.

West and North Yorkshire

Allied Colloids Ltd, Low Moor, Bradford. **British Gas Corporation (NE Gas)**, Tingley, Leeds. **Elida Gibbs Ltd**, Wakefield, West Yorks. **Flogas Ltd**, Cross Green Industrial Estate, Leeds. **Imperial Chemical Industries plc**, (ICI Organics Division), Huddersfield, West Yorks. **Hickson & Welch Ltd**, Castleford, West Yorks. **A. H. Marks & Co. Ltd**, Wyke, Bradford. **Melmerby Industrial Estate Ltd**, Ripon, North Yorks. **J. Revis & Sons**, York, North Yorks.

Wales

Amoco (UK) Ltd, Milford Haven, Dyfed. **Anglesey Aluminium Metal Co. Ltd**, Holyhead, Gwynedd. **Associated Octel Co. Ltd**, Amlwch, Gwynedd. **BOC Ltd**, Newport, Gwent. **BP Angle Bay Ocean Terminal**, Pembroke, Dyfed. **BP Chemicals Ltd**, Port Talbot, West Glamorgan. **BP Oil Ltd**, Neath, West Glamorgan. **BP Oil Ltd**, (Llandarcy Refinery), Swansea, West Glmorgan. **British Gas Corporation (Wales Gas)**, Wrexham, Clywd. **British Gas Corporation (Wales Gas)**, Llandudno, Gwynedd. **British Gas Corporation (E. M. Edwards)**, Neath, West Glamorgan. **British Gas**, Merthyr Tydfil, Mid Glamorgan. **Calor Gas Ltd**, Neath, West Glamorgan. **Consoldiated Beryllium Ltd**, Milford Haven, Dyfed. **Gulf Oil Refining Ltd**, Mildford Haven, Dyfed. **Imperial Chemical Industries plc**, Sully, South Glamorgan. **Inco Europe Ltd**, Clydach, Swansea. **Monsanto Chemicals Ltd**, Ruabon, Clywd.

Monsanto Chemicals Ltd, Newport, Gwent. **Pembroke Cracking Co.** Pembroke, Dyfed. **Shell UK Oil**, Amlwch, Gwynedd. **Texaco Ltd**, Pembroke, Dyfed. **UKF Fertilisers Ltd**, Carmarthen, Dyfed.

Scotland

Borg Warner Chemicals Ltd, Grangemouth, Central Region. **BP Chemicals**, Grangemouth, Central Region. **BP Developments Ltd**, Lerwick, Shetland. **BP Oil Ltd**, Grangemouth, Central Region. **BP Oil (Dock Storage) Ltd**, Grangemouth, Central Region. **BP Oil, Grangemouth Refinery Ltd**, Garelochhead, Helensburgh. **BP Oil, Grangemouth Refinery Ltd**, South Queensferry, Lothian. **British Gas Corporation**, Peterhead, Grampian. **British Gas Corporation (Production and Supply)**, Airdrie, Strathclyde. **Britoil**, Nigg, Highland Region. **Calor Gas Ltd**, Grangemouth, Central Region. **Calor Gas Ltd**, Fallin, Stirling. **EniChem Elastomers Ltd**, Grangemouth, Central Region. **Essochem Olefins Inc**, Burntisland, Fife. **Essochem Olefins Inc**, Cowdenbeath, Fife. **Imperial Chemical Industries plc**, (ICI Organics Division) Grangemouth, Central Region. **Nobels Explosives Co. Ltd**, Irvine, Strathclyde. **Norsk Hydro Fertilisers Ltd**, Leith, Edinburgh. **Occidental of Britain Inc**, Flotta, Orkney. **Rohm & Haas (UK) Ltd**, Grangemouth, Central Region. **Scottish Agricultural Industries plc**, Leith, Edinburgh. **Scottish Gas Provan Works**, Provan, Glasgow. **Scottish Gas**, Stornoway. **Scottish Gas**, Edinburgh. **Shell UK Exploration & Production**, Peterhead, Grampian. **Shell UK Exploration & Production**, Burntisland, Fife. **Shell UK Exploration & Production**, Cowdenbeath, Fife. **G. R. Stein Refractories Ltd**, Linlithgow, Lothian. **Thomas Tait & Sons Ltd**, Inverurie, Grampian. **Total Oil Marine Ltd**, Peterhead, Grampian.

NOTES

EDITOR'S NOTE

In the original version of this book the author placed a '*sic*' after each occurrence of a masculine noun in a quotation. These included the following: man (in the sense of humanity), man-machine, manning, man-hours, watchman, he (worker), him, etc. Although the editor felt that it was worth noting the sexist character of the language in those quotations, he thought it excessive to proliferate '*sic*'s especially because the book does not discuss gender relations. Moreover, in many cases the masculine term is accurate as chemical process workers are mostly male; toxic hazards are often considered more acceptable for men than for (child-bearing) women, and this perceived difference itself warrants a detailed political analysis.

INTRODUCTION

1. This is a paraphrase from the song by Jack Warshaw, 'If They Come in the Morning', about police terror.
2. For an indication of the growth of chemical production in the USA, see M. S. Legator and S. J. Rinkus, 'The chemical environment and mutagenesis', in Society of Occupational and Environmental Health, *Work and the Health of Women*, New York: SOEH, 1979, pp. 6–8, 21–3. See also H. D. Crone, *Chemicals and Society: A Guide to the New Chemical Age*, Cambridge University Press, 1986, pp. 14–17. For world-wide

growth of plastics production see H. Vainio *et al*, 'Chemical hazards in the plastics industry', in *Occupational Cancer and Carcinogenesis*, Washington, Hemisphere Publishing Corporation, 1981, p. 260.

3. APPEN (p. 174) has reprinted claims by the Indian radical science group Kerala Sastra Salutya Parishad that the Bhopal plant had originally been rejected by Canada before being shifted to India.

4. The Minister for Labour in the Madhya Pradesh state government, Shyam Sunder Patidar, resigned on 28 December 1984 after accepting moral responsibility for the disaster. Two Labour Ministry officials were suspended and two others sent on leave (*Guardian*, 29 December 1986, p. 4). The same day the chief minister of Madhya Pradesh state, Arjun Singh, announced he had dismissed C. P. Tyagi, the chief inspector of factories, who 'went on renewing the licence of the Union Carbide factory every year without taking into cognizance the safety lapses in the factory' (Singh quoted in *New York Times*, 29 December 1984, p. 4).

5. The *Financial Times* (10 December 1984) continued, 'So little is known about [MIC's] lethal capacity that Union Carbide India Ltd assumed there would only be a small loss of life. Even 1–2 hours later a maximum figure of under 100 deaths was expected.' Nor can this ignorance be shrugged off by referring to the poor quality of Indian management: 'Tear-gas' Mukund, for example, works manager at Bhopal, has an engineering degree from Cambridge and did postgraduate work at the Massachusetts Institute of Technology.

6. One study found 74 percent of those in affected areas fled on foot, 6 percent on a vehicle (motorized or bicycle) and 21 percent stayed. None of those who went on a vehicle died. A study by the Centre for Social Medicine, Jawaharlal Nehru University, New Delhi, found that 'those who died are the poorest. More than half the affected people belong to an income group – about 150 rupees per head per month – which cannot afford two full meals a day around the year. Those who died were even more disadvantaged than the overall affected population.' Henry Falk of the US Centers For Disease Control noted that people who were lucky (!) enough to live in well-sealed homes managed to avoid the choking gas, while some escaped by climbing to a higher level.

Chapter 1

1. *The picture that emerges from conversations with some workers and from the government's report is that Ashraf Khan had opened the flanges of the pipeline for carrying out maintenance work and had inhaled some amount of liquid phosgene when he removed the breathing mask after completing his job. He died a day later. According to the workers whom this writer met, when Ashraf Khan began to work on the phosgene*

pipeline, the head of the production department had not certified, as the safety rules stipulate, whether the pipeline had been emptied of phosgene or not. The workers alleged that it was this disregard for even basic safety procedures on the part of the head of the production department that had caused Ashraf Khan's death.

*Instead of finding out how liquid phosgene had managed to seep into the pipeline in the first place since (a) the MIC plant was shut down for maintenance, and, therefore, phosgene production had completely stopped, and (b) according to the company's rules, phosgene is consumed immediately after its manufacture and is never stored, even in small quantities, the government's report seeks to shift the blame for the accident on carelessness on the part of workers. (*Economic and Political Weekly, *15 December 1984, p. 2110)*

This Indian account should be compared with the following UCC account: 'In an accident three years ago, a Bhopal worker "panicked" when phosgene gas leaked at the plant, Carbide disclosed last week. According to a Carbide spokesman, two Bhopal plant workers were disassembling a valve when the gas started leaking. One worker panicked and removed his protective hood, according to the spokesman, and died the next day' (*Chemical Marketing Reporter*, 10 December 1984, p. 53).

2. For further discussion on 'safe' and threshold levels see Epstein (pp. 64–7, 307–8), Kinnersley (pp. 116–18), Fishbein and 'The threshold controversy' by Dr P. Gehring of Dow Laboratories (in *New Scientist*, 18 August 1977, pp. 426–8). For a classic analysis of the political use of these levels in the Three Mile Island disaster, see Levidow (1979).

3. These establishments have not been named by the British Health and Safety Executive. An educated guess would include the British chemical warfare research station at Porton Down. It is hard to imagine what other type of research establishment would have a use for an intermediate chemical like MIC. At the time Ciba-Geigy were reported to be the only company in the UK using MIC, but in 1986 the *Guardian* (28 August 1986, p. 30) reported a spill of MIC from a drum on open land at the Clay Cross Company in Clay Cross, Derbyshire. The factory was evacuated and fifteen fire-fighters received check-ups at hospital.

4. According to his 1974 publication, Dr Ballantyne's qualifications are B.Sc, MD, Ph.D, MRC Path., FI Biol. Dr Ballantyne has edited *Forensic Toxicology: Proceedings of a Symposium held at the Chemical Defence Establishment, Porton Down, 29–30 June 1972* (Bristol: John Wright & Sons, 1974). Dr Ballantyne is listed as 'Senior Medical Officer (Research), Ministry of Defence'. Dr Ballantyne contributed one paper to the above volume: 'The forensic diagnosis of acute cyanide poisoning' (pp. 99–113). Other topics covered in the volume include drug detection, post-mortem levels of drugs, paraquat, carbon monoxide, lead and mercury. For the second of Ballantyne's volumes

which I was able to obtain, he is listed to UCC in Charleston, West Virginia, but his co-editor is Paul H. Schwabe of the Chemical Defence Establishment, Porton Down, Wiltshire SP4 0JQ. This volume, *Respiratory Protection: Principles and Applications*, was published by Chapman & Hall in 1981. Ballantyne is also listed as Adjunct Professor of Toxicology at the University of Pittsburgh in this volume.

Of the volume's twenty-three authors, no fewer than six are listed as from Porton Down, while one is from the Personnel Research Establishment of the British Army. This array of experts is complemented by three authors from British Nuclear Fuels Ltd and one from the asbestos firm Turner & Newall. Dr Ballantyne's contributions are 'Occupational health aspects of respiratory protection' (chapter 6), 'Inhalation hazards of fire' (chapter 20) and 'Toxic effects of inhaled material' (chapter 15). A bibliography on page 130 notes Dr Ballantyne was preparing in 1980 a paper entitled 'The acute toxicity of hydrogen cyanide and its alkali salts'. Dr Ballantyne has also edited *Current Approaches in Toxicology*, published in Bristol by Wright in 1977.

5. Col. K. M. Rao and Dr R. K. Srivastava, Defence Research and Development Establishment, Gwalior, were present at the Indian Council on Medical Research review meeting on the health effects of exposure to the toxic gas at Bhopal held in Bhopal on 3–4 May 1985, while Lt-Col. Sharma was at the third meeting of the working group on sodium thiosulphate held at the Indian Council on Medical Research headquarters on 4 April 1985. In an interview held in August 1985 Dr Sadgopal of the Morcha spoke of how an electrode intended for thiocyanate-level measurement, which was necessary for work in Bhopal, was originally intended 'to go to some lab in Gwalior' (APPEN, p. 49).

6. It's touching that later again a revealing interview in the *New York Times* business pages (19 May 1985, section 3, pp. 1, 8) was to let us know of Mr Anderson's difficulty in sleeping after Bhopal; this revelation was spread over two pages in a newspaper that had covered the actual health conditions of the victims of Mr Anderson's company only once in the previous few months.

7. That many of the recipients of these medals were Hindus or Muslims only adds to the irony. Mother Teresa's Missionaries of Charity were reported to be operating a treatment camp at the railway station of Bhopal for UCIL (*Chemical and Engineering News*, 11 February, p. 19). Later, that wonderful leader of the Roman Catholic Church, Karol Jozef Woytyla, also known as John Paul II, took up forceful suggestions that his speeches during his tour of India should be made more relevant to the reality of life on the subcontinent; he described Bhopal as a 'sad event' resulting from 'man's efforts to make progress' (*Irish Times*, 7 February 1986, p. 4).

8. For sources on this issue see Holmes (1982) and Mac Sheoin (1985, 1987). The most hilarious industry response to the double-standards charge was reported by W. Worthy in *Chemical and Engineering News* (6 January 1986, p. 16): 'As to the last point all the

chemical companies contacted deny that they ever have had different safety or operating standards for US and foreign operations. However, they add, they're going to try to do even better in that respect.' This translates as: We didn't do it, honest, but we'll not do it even more in the future. Note also that the ambiguous word 'standards' creeps in here. Farther down the same page, Monsanto's O'Neill is quoted as saying: 'Prior to Bhopal, our headquarters safety team would do comprehensive safety audits of US and European plants every year, but only every two years in other parts of the world; that is, in Africa and Asia.' These aren't different standards, though.

9. Of course, it's not like this in metropolitan countries. In Livingston New Town, Scotland, where UCC proposed to build a gas-mixing plant, people would have been living 500 yards from the plant (*Financial Times*, 8 February 1985). In Béziers as well, people live only a few hundred yards from a UCC subsidiary plant that used MIC (*Wall Street Journal Europe*, 20 December 1984, p. 1). A similar situation exists in Institute, West Virginia. Some 75 percent of the US population have been reported to be within striking distance of a chemical plant.

10. Perhaps UCC's most heinous deed was the Gauley Bridge killing. 'In a Depression-era scandal in West Virginia, 476 workers died of the lung disease silicosis while building a diversion tunnel for a Carbide-sponsored hydroelectric project. Hundreds of others suffered and died later from the disease, in what labour experts call the worst incident of its kind in US history' (*Wall Street Journal Europe*, 3 January 1985, p. 2). UCC was until recently a major producer of weapons of mass murder for the US military machine: this grew from UCC's key role in the Manhattan Project. Some indication of working conditions at UCC's uranium enrichment plant at Paducah, Kentucky, is provided by the following posthumous testimony presented at the Citizens' Hearings on Radiation Victims in April 1980:

There was uranium dust all over the floor, so deep that you could see the footprints in it. We wore film badges to monitor our radiation dose; they were collected once a month and sent to Oak Ridge Laboratory, also run by Union Carbide's Nuclear Division, for analysis, but we never heard the results. One time, three or four of us laid our film badges on top of a smoking chuck of uranium for eight hours. Still we didn't hear from them.

Further sources on UCC's past criminal record are listed in Mac Sheoin (1986, pp. 449-50). See also Agarwal (1985, pp. 18-23), and *Our Own Worst Enemy: The Impact of Military Production on the Upper South* (New Market, Tennessee: Highlander Research Center, 1983). For the Gauley Bridge killing see M. Cherniak, *The Hawk's Nest Incident* (Yale University Press, 1987), and *Health/PAC Bulletin* (November/December 1977, pp. 9-16).

11. The sources for these quotes are *European Chemical News* (1 April 1985, p. 14) and

Newsweek (1 April 1985, p. 40). Interestingly, the industry-press coverage emphasized the questions UCC's report left unanswered.

12. This particular trick is not confined to UCC. The *Asia Labour Monitor* (September 1983, p. 4) reported that six members of the union work-site committee at the Chemical Company of Malaysia's Padang Jawa plant were suspended for five days without pay and the committee's secretary was sacked for writing to the government complaining of the factory's pollution problems. The Chemical Company of Malaysia is a subsidiary of the highly ethical British multinational, Imperial Chemical Industries.

13. This accident confirms Nichols and Armstrong's (1973) conclusions that accidents occur most often when workers are attempting to restart production. See also Chapter 8, note 11.

14. Developing 'resistance' in this manner is a common management suggestion in a wide range of industries using toxic chemicals.

15. This *Business India* article (anon., 1986) presents some marvellous examples of the knots UCC management tied itself into in its desire to avoid liability and should be read in its entirety.

16. This narrowing of the definition of the accident avoids questions such as why the refrigeration system was turned off, why the emergency relief system was too small, why staff levels had been cut, why no qualified engineer was on site, why instrument readings were untrustworthy and why such a massive amount of highly toxic chemicals was stored on site. It implies that the addition of the water was the sole cause of the accident rather than one cause among many. Note the similarities in the treatment of Ashraf Khan's death (note 1 above). See also the discussion in Chapter 8, 'Accidents will Happen'.

17. This is a standard industry ploy. A similar example of this tactic came during the Irish High Court case between Irish farmer John Hanrahan and Merck, Sharp & Dohme in 1985. Hanrahan claimed emissions from Merck, Sharp & Dohme's nearby factory had poisoned his cattle and his family. Defence witnesses, from Irish Department of Agriculture officials to farm management consultants, argued that the 'lack of thrift' (illnesses) in Mr Hanrahan's cattle could be explained from a retrospective analysis of such 'production parameters' as fertiliser use and cattle density, despite the fact that local officials had dismissed such factors during their practical examination of the farm. The Merck, Sharp & Dohme witnesses claimed their retrospective theoretical analysis had more validity than direct observation of the farm during the period in question. This was a case of a theoretical model being given precedence over reported reality due to its allegedly more 'scientific' nature.

18. Central Bureau of Investigation officials who

had disconnected the pipeline from the storage tank on the morning after the leak, drained out as much as 27 litres of water from the structures on 14 February. This totally refutes the theory of sabotage – with water deliberately poured into the tank – being floated by certain company officials. The water was found in almost all connecting pipelines tracing the entire route from the point of washing to the Relief Valve Vent Header. (Sunday, 7–13 April 1985, p. 19)

For an allegedly scientific account of the accident, it is strange that UCC's report omitted mention of a jumper line from the pressure-vent header to the relief-valve-vent header, one of the major possible routes for water 'inadvertently' to enter Tank 610. UCC finally admitted its existence only following publication of the international trade-union report on Bhopal; even then UCC claimed it to be a modification of the original design which was undertaken by UCIL.

19. Similarly the *Journal of the American Medical Association* (12 April 1985, p. 2013) reported Warren Anderson as denying sabotage should be inferred from the UCC statement that water was put in the tank 'inadvertently or deliberately'. UCC was having it both ways again. Its memo of law in support of its motion for dismissal of the US law cases cites the possibility of sabotage four times. (This document was dated 31 July 1985.) Of seven critical issues UCC enumerated, the fourth was: 'Was it caused by the deliberate actions of an Indian saboteur?' (APPEN, p. 199). The first paragraph of its presentation on the background of the US suits is as follows:

These actions arise out of the release of methyl isocyanate from a plant owned and operated by Union Carbide India, Limited in Bhopal on 3 December 1984. Union Carbide does not know if the cause of the release was accidental or the result of a deliberate act. A group of Indian Sikh extremists, which call itself 'Black June', has claimed responsibility for the incident. (APPEN, p. 200)

Later still the submission on behalf of UCC reads: 'Was it an accident or was it, as the available evidence information increasingly indicates, caused by a deliberate act of sabotage?' (APPEN, p. 207). Alas, as usual, this available information was available only to UCC.

Interestingly, as late as November 1985, UCC's counsel, Bud Holman, was admitting he had no direct evidence of sabotage 'but says that getting the water into the tank "was so improbably inadvertent that our conclusion is that it was deliberate". He adds that the plant library contained information that would have been useful to a saboteur and that it was open to most employees' (*Chemical and Engineering News*, 2 December 1985, p. 31). This conflicts with reports that plant workers had difficulty in getting access to even the MIC plant operating manual.

20. It's only a small point, but UCC interviewed no eyewitnesses after the leak. According to Ron Van Mynen of UCC, 'We do not have sufficient information to

comment on specific actions that particular individuals may have taken or not taken. To do that would have required immediate and searching interviews with people on the scene during the incident, and we did not have the opportunity to conduct such interviews' (Van Mynen, 1985, p. 2).

Chapter 2

1. The *Guardian* (7 December 1984, p. 12) reports estimates of deaths in the Ixhuatapec killing as high as 3,000. Estimates of deaths in these cases are always affected by lack of baseline. Thus the Vila Scola shanty town in Cubatão, Brazil, which was devastated in a fire caused by leaking petrol in February 1984, had either 8,000 or 12,000 residents. No one can be certain, as the government never documented the population there. Constant migration from the countryside has also to be considered. Local officials said there are normally some 60,000 migrant labourers in Bhopal: these normally have no official residence nor family (Everest, 1985, p. 61).

2. A charge is defined as 'in a proft-and-loss account, a deduction from revenue' by D. French, *Dictionary of Accounting Terms*, Institute of Chartered Accountants, 1985.

3. Operation Faith gave the company output estimated at 6 million rupees, while the alternative would have involved costs of 1 rupee per cubic foot of nitrogen, without considering the price of the massive quantities of caustic soda required (*Economic and Political Weekly*, 12 January 1985, p. 57).

4. *Chemical and Engineering News* (11 February 1985, pp. 23–4) gives a marvellous *Boys' Own*-type description of how India's scientists and technologists rallied to Varadarajan in those troubled days after the gas leak.

5. Non-appearance of such reports is no less common in India than in other countries. As one example of this, the *Guardian* (31 October 1986, p. 8) reported the non-appearance of a report promised by the Rajiv Gandhi government on the pogroms of Sikhs following the assassination of Indira Gandhi. Shrivastava (1987, p. 65) reports: 'Mass deaths in religious and caste riots in India are routinely under-reported by government sources in order to prevent conflicts from escalating. For example, in the 1984 Punjab riots involving Hindus and Sikhs, the government reported 2,000 deaths, while press reports and non-governmental investigations put the figure at between 8,000 and 10,000.'

6. A different tactic was used by UCC in the USA. 'In Institute we were simply refused permission to film at all' (*Listener*, 6 June 1985, p. 7). So much for UCC's fabled openness with the press.

7. Private communication, Satingath Saringu, 18 November 1986.

8. The major source for this account is the Delhi Committee on Bhopal Gas Tragedy (1985). As one example of the divisions between groups organizing around Bhopal, the

National Campaign Committee on Bhopal, associated with the Morcha, called for this report to be withdrawn at its meeting on 30 August/1 September 1985 (APPEN, p. 64).
9. One example of the racist treatment of the Bhopal aftermath in the Western press is in the story of David Bergman, which received constant coverage in the *Guardian* and *The Times* while the arrests of the Bhopal Group for Information and Action members were ignored. Similarly, the mass arrest mentioned merited only a one-paragraph news note in the British press. No disrespect is intended to David Bergman, who himself drew attention to this racism:

It is unfortunate that my fate has been publicized much more than that of the victims of the gas disaster. Here I am just because I am British, yet there are hundreds of thousands of victims in the poorest conditions who do not have the opportunity that I have and who are continuing to suffer in a very devastating manner. (Guardian, 2 January 1987, p. 2)

Chapter 3

1. Arjun Singh was reportedly so close to UCIL that he permanently retained two rooms at the company's luxurious guest-house in the hills above Bhopal to entertain his associates.
2. The opposition's call for the use of ultrasound and amniocentesis reveals a rather naïve and uncritical approach to high-technology medical fixes. Use of ultrasonography has been criticized by feminists and other health activists because of the lack of information on its possible long-term effects, while amniocentesis has been promoted without its limitations being made clear. (For some discussion on this, see C. Norwood, *At Highest Risk*, New York: Penguin, 1981, pp. 81–6 and *The Custom-Made Child? Woman-Centred Perspectives*, Clifton, NJ: Humana, 1981, pp. 105–15. Other aspects of these technologies, such as the eugenic ideology that accompanies them, are discussed in the critical feminist literature on the new reproductive technologies.) A similarly uncritical attitude to sodium thiosulphate treatment led some opposition forces to a position where it seemed that all that was required to deal with the killing's medical aftermath was the use of a particular drug/antidote – a technical fix if ever there was one.
3. A comparison with Seveso is instructive here:

... one must also remember the start of the drama. Between 10 and 12 July 1976, while the technical director of Givaudin immediately thought of the terrible hypothesis of the dioxin, from 11 July the company was satisfied with the admission of the escape of a 'cloud of herbicide'. It took ten days of going to and from Geneva, since the company had no laboratory on the spot, for Hoffman LaRoche, after having obtained the proofs requested, to admit the presence of the poison (20 July) ... Roche would not have admitted the

presence of the poison unless the Milan experts had uncovered the situation. (Lagadec, 1982, p. 332)

4. The *British National Formulary* (1981, no. 1) notes on page 24: 'A number of preparations are available for use as an aid in the removal of poisons other than those mentioned above, i.e. cyanides (dicobalt edetate, sodium nitrite and thiosulphate) . . .'

Chapter 4

1. A. H. Hermann, legal correspondent of the *Financial Times*, noted (28 March 1985, p. 13):

Finally the Indian government would be spared the embarrassment of having the dirty linen of its local and central administration laundered in US courts, where the Union Carbide lawyers would be bound to establish that the goverment had at least partial responsibility for the tragedy. This concern must have played a role in the swift legislation by which the Indian government assumed powers to sue Union Carbide for compensation on behalf of the victims and in this way to establish for itself a position in the negotiations for a settlement.

2. Holman is a partner in Kelley, Drye & Warren of New York. He had previously done defence work on Agent Orange and asbestos and reportedly had done product liability work for UCC (*American Lawyer*, November 1985, p. 57).

3. The Delhi Committee report alleges the local Red Cross activities are intended to benefit UCC:

Dr Nagu, director of health services in Bhopal, has a relative, another Nagu, who is responsible for security of Carbide plants in India. This person is also the chairman of the local Red Cross unit, which is significant because Warren Anderson is on the board of the Red Cross in America. Red Cross has secured three clinics in Bhopal to treat the gas victims, and local activists allege that medical records from these are being sent to Union Carbide. (1985, p. 12)

Similar allegations were later reported by the Bhopal Group for Information and Action in their newsletter, *Bhopal* (October 1986, p. 8):

The Roman Catholic missionary hospital Agape is alleged to have passed on records to UCC which can be used to prove that effects of MIC on victims were minimal. Expensive equipment was provided by UCC, and the salaries of doctors and other expenses were also borne by the multinational firm. The hospital reportedly examined 50,000 patients and confidential medical records were meticulously prepared. Medical reports, it is learnt, were 'doctored' to suit the interests of UCC in the legal case.

On the basis of medical reports, AGAPE concluded that the 'gas victims' were suffering, not from MIC inhalation, but from various pre-existing problems like undernourishment. According to AGAPE, most of the patients were living in TB-endemic areas, and were suffering from TB and asthma even before the gas leak. Inhaling MIC merely worsened their condition.

UCC provided expensive equipment to the Central Railway Hospital in Bhopal, and funded the Red Cross clinics and the Eye Hospital in Bhopal, all of which prepared similar reports after examining a large number of patients. The closure of the missionary hospital soon after the Union (i.e. central) government filed suits for compensation in Bhopal has raised doubts in medical circles. Evidently, gas victims were attracted to the hospital by the free rations and edible oils offered, but had to undergo medical tests in return.

4. The situation has always been thus. In his study of industrial hazards in the USA at the beginning of the twentieth century, Gersuny notes:

In the adversary process spelled out under the employers' liability laws, the impoverished victims of industrial injuries had to rely on the services of attorneys practising on a contingent-free basis. While these practitioners who thus served the poor 'were denounced by the professional élite as inferior ambulance chasers or shysters', they provided virtually the only available legal representation that could be gotten by injured workers. (1981, p. 54)

Chapter 5

1. The quotation comes from A. Szasz, 'The process and significance of political scandals: a comparison of Watergate and the "Sewergate" episode at the Environmental Protection Agency', *Social Problems*, February 1986, pp. 202–17. For sources on the US experience of deregulation see T. Mac Sheoin, 'The dismantling of US health and safety regulations under the first Reagan administration: a bibliography', *International Journal of Health Services*, vol. 15, no. 4, pp. 585–608, 1985. For deregulation in Britain see GMBATU, *The Freedom to Kill? A Response to Deregulation in the Workplace*, Esher, Surrey, 1985.

2. This was echoed in the Western response to the accident at the nuclear-power reactor in Chernobyl. As one example, Peter Walker, Energy Secretary, when announcing the British government's decision to build a pressurized water reactor at Sizewell in Suffolk, said that British reactors had a 'superior safety culture' to that at Chernobyl (*Guardian*, 13 March 1987, p. 6).

3. For criticism of the British implementation of the Seveso directive, see *Health & Safety at Work*, May 1986, pp. 37–47. For a useful examination of the implementation of the public-information provisions of the Seveso directive, see B. Wynne, 'Risk communication for chemical plant hazards in the European Community "Seveso"

directive – some observations based on comparative empirical studies', draft paper for presentation at the annual meeting of the Society for Risk Analysis, Boston, 11 November 1986.

4. For a similar chorus by European chemical firms, see *European Chemical News*, 11 March 1985, pp. 10, 12.

5. For coverage of the Clean Sites initiative, see *Business Week*, 4 June 1984, p. 41, and *New York Times*, 14 April 1984, p. 7, and 1 June 1984, p. A14.

6. Given the industry's previous failure to share information with employees this promise bodes ill for the public. See Epstein (1978) and Frankel (1981); also P. Broseur, *Outrageous Misconduct*, New York: Pantheon, 1986.

Chapter 6

1. The implication here is that those involved are either hypochondriacs or professional litigants. This is a common allegation. In the case of the pollution controversy over the Merck, Sharp & Dohme factory in Ballydine, Co. Tipperary, Ireland:

One of the more affected farmers found himself the subject of rumours of a diverse and persistent nature. These have been repeated to us in the course of our inquiries. They include suggestions of bad farm management, that his silage is contaminated, that his cattle have diseases, that he is a 'professional litigant' and/or 'gone in the head'. Some of these rumours emanated from Limerick and Dublin and had currency among persons who were not sure where exactly Ballydine was located. Rumours in themselves are of no interest, except in a case where they persist and where one succeeds another with uncanny and coincidental variety! (Making Their Merck on Ireland, *Dublin: Merck, Sharp & Dohme Study Group, 1982, pp. 22–3)*

2. Whatever aldicarb oxime and methylene chloride are, they are neither pencils nor perfume. Aldicarb oxime is known primarily as an eye irritant, but few studies have been done on its health effects: there is controversy over whether or not it causes cancer. Methylene chloride can cause central nervous system, respiratory, liver and blood disorders as well as eye irritation. Animal tests have shown it can cause cancer.

3. Similarly, while the media were highly reassuring about the safety record of the Institute plant, a citizen's report stated, 'However, residents in the small, predominantly black town which crowds close to the plant, remembered frightening incidents at the plant, including an explosion and evacuation of the community in June 1954. Workers recalled a series of fires, explosions and leaks, and their own exposure to MIC gas' (Agarwal, 1985, p. 11).

4. For a technical criticism, see D. J. McNaughton *et al.*, 'Evaluating emergency

response models for the chemical industry', *Chemical Engineering Progress*, January 1987, pp. 46–51.

5. This concern was also possibly related to the Seveso disaster where a runaway reaction had shown the non-existence of an emergency-relief system in that plant. For background on the Design Institute for Emergency Relief Systems, see *Chemical Engineering Progress*, June 1983, pp. 11–12. MacKenzie (1986) is a good source for the problems in this area.

Chapter 7

1. See for example the pages on which the following toxic incidents in the USA were reported in the *New York Times:*
Two thousand five hundred people evacuated from Little Rock, Arkansas, after leak of ethylene oxide from railway tanker (p. A12, 2 January 1985)
Fourteen injured and 1,500 evacuated following leak of hydrofluoric acid from railway tank wagon (p. A14, 5 February 1985)
Three dead and 2 injured in fire involving toxic gases at Singer Furniture Co. (p. A10, 19 February 1985)
Fire and explosion at Hercules plant which killed one person and injured another (p. 26, 23 February 1985)
Five deaths from explosion of natural-gas pipeline (p. A16, 30 April 1985)
One thousand people evacuated after explosion in propane tank wagon (p. 7, 27 May 1985)
Three workers burnt in explosion at Glow Plastic Corp. plant (p. B8, 31 May 1985)
Twenty-one deaths following explosion at Aerflex Corp. fireworks plant (p. A16, 26 June 1985)
One thousand people evacuated after spill and fire at Wilbur-Ellis Co. pesticide plant (p. A16, 27 June 1985)
Ten thousand people evacuated following fire spewing out clouds of gas containing hydrogen chloride from sewage plant (p. A14, 17 July 1985)
Eleven fire-fighters injured and houses evacuated following fire at plant storing cyanide compounds (p. B3, 15 August 1985)
Four hundred people evacuated following vinyl acetate leak from tank wagon (p. A12, 19 August 1985)

2. For background on the crisis in the US chemical industry and the strategic responses to it, see J. Bower, *When Markets Quake*, Boston: Harvard Business School Press, 1986.

3. See S. Epstein, 'Cancer and inflation', *The Ecologist*, October/November 1979,

pp. 236–46, and T. Anthony, 'Environment and inflation', *Environmental Science and Technology*, April 1979, pp. 400–4.

4. For one front-line example of this continuing battle over the causes of ill-health, see Suser's (1985) superb report on UCC's operations in Yabucoa, Puerto Rico.

5. This was originally intended as a sardonic jest. S. Dube reported in early 1987:

The EPA derailed discussion of a 1985 Congressional bill that would have required strong best-available technology use on industrial facilities by arguing that area sources were a greater health risk to communities and should be regulated first ... According to Congressman Henry Waxman (Democrat, California), co-author of the bill and chairman of the subcommittee, 'for the EPA to insist that wood-stoves and dry-cleaners are the main sources of our toxic-air pollution problem is incredible. They simply ignore the chemical plants around the country, which pose a serious danger to people who live near by, suffer ill health, and are dying because of exposure to these chemicals.' (Science for the People, January/February 1987, p. 22)

A similar line has been advanced by the US Food and Drug Administration. The US government's top food-safety official said US consumers worry too much about chemicals in their food, when they should be more concerned about disease-causing microbes in food. (*Chemical Marketing Reporter*, 24 November 1986, p.7)

6. See *Newsweek*, 5 May 1986, p. 53. See also the reasons why ICI are not pushing research on tamoxifen as an anti-cancer drug, *Guardian*, 4 June 1986, p. 23.

Chapter 8

1. For industry's presentation of its constant drive to increase safety, see for example J. De Mott, 'An unending search for safety', *Time*, 17 December 1984, p. 21.

2. Lagadec (p. 175) gives a 1980 version of these figures:

Number of Cases

Decade	Number of cases	Total loss ($ millions)	Average loss ($ millions)
1950–9	7	173	24.7
1960–9	16	404	25.2
1970–9	46	1318	28.6

Shrivastava notes:
There is no question that the twentieth century has produced a rapid rise in the rate of major industrial accidents and deaths. Half of the century's twenty-eight most serious industrial accidents (those killing at least fifty people) have occurred since 1977. Three of these disasters, including the Bhopal disaster, occurred in 1984 alone – all in rapidly industrializing countries ... Evacuations caused by industrial accidents follow a similar

pattern. Of the 919,000 people evacuated in major rescue efforts (2,000 or more people) between 1967 and 1984, more than 90 percent were evacuated in incidents that have taken place since 1979. More than 50 percent of that figure were evacuated in the year 1984. (1987, p. 9)

3. See the review of Perrow's *Normal Accidents* in *Chemical and Engineering News*, 29 July 1985, pp. 30–1, which concentrates on the nuclear power question and ignores Perrow's basic thesis. Yet Perrow's book is reasonably kind to the chemical industry. He declares (p. 114) vapour clouds to be 'fairly exotic'. Yet the article on vapour clouds he cites (Davenport, 1977) lists 230 dead and 1,375 injured in vapour-cloud incidents from 1921 to 1976. (The early period has few incidents listed.) A British listing of vapour-cloud incidents from 1921 to 1978 (Slater, 1978) gives 826 dead and 5,161 injured. Not bad for a fairly exotic type of accident.

4. Lagadec quotes from the official report on Flixborough:

We completely absolve everyone of any grievance according to which their desire to restart production could have led them, in full knowledge of the facts, to get a dangerous process going without paying attention to the safety of those who did the work. We have, however, no doubt that it was indeed this desire which led them to neglect the fact that it was potentially dangerous to restart production without having examined the remaining reactors and having determined the cause of the fissure in the fifth reactor. In the same way, we have no doubt that the error in appreciation as concerns the problem of connecting reactor No. 4 to reactor No. 5 was largely due to the same desire. (1982, p. 357)

5. According to the report by G. Thyagarajan of the Indian Council for Scientific and Industrial Research, 40 percent of all chemical accidents in India are due to design faults (*Chemical and Engineering News*, 9 February 1987, p. 5). Similarly, faulty design was to blame for the killing at Ixhuatapec, Mexico, the month before Bhopal; see M. Samuelson, 'Design blamed for Pemex blasts', *Financial Times*, 16 July 1985, and F. Pearce, 'After Bhopal, who remembered Ixhuatapec?', *New Scientist*, 18 July 1985, pp. 22–3.

6. An observation by Lagadec is relevant here:

One must also consider the great difficulties encountered in the field of product knowledge, of reactions caused and possible after-reactions. Despite laboratory studies, pilot studies, half-size, and years of research (between five and seven years for a petrochemical process), essential points of safety remain obscured. In the case of an explosive toxic product, for instance, a safey analysis was requested because of extremely serious dangers which the installation apparently causes to the neighbouring agglomeration. One became aware of the difficulties of the work. When one wanted to define the means of prevention, one asked for indicators which would raise an alert when a dangerous reaction started; when one wanted to define the means of intervention to be set up, one asked about the time one would have to remove and neutralize the product that was not yet in the reaction process . . . despite mathematical studies, laboratory studies, document studies one hits on something unknown that was difficult to solve: the kinetics of the reaction in question are not well understood. (1982, p. 307)

The position was put more baldly by a process development manager in Scotland in 1978: 'The problem is that we [who were responsible for introducing new technology] are trying to introduce chemical *engineering* into a chemical works where the dominant pattern of manufacturing is still closer to alchemy than anything else' (Buchanan and Bessant, 1985, p. 298).

7. *The capitalist system can only maintain itself by trying to reduce workers into mere order-takers, into 'executants' of decisions taken elsewhere. At the same time the system can only function as long as this reduction is never achieved. Capitalism is constantly obliged to solicit the participation of workers in the process of production . . . On the other hand capitalism constantly has to* limit *this participation. (Cardan, 1969, p. 45)*

8. The ergonomics approach, which sees chemical-plant workers' main job as monitoring and treats it as watch-keeping, ignores the actual working conditions. These specialists inappropriately use watch-keeping studies, which were originally developed in the military use of radar. This military carry-over is not surprising, as ergonomics arose from 'the need to put high-performance human operators in control of the new and complex military systems that were rapidly developed at the outbreak of the Second World War' (Human Reliability Assessment Group, 1985). Other elements of toxic capital's reassurance programme also have military origins: 'The present subject of reliability owes much of its development to such pioneers as Robert Lusser, who in Germany in the early 1940s in the missile field found increasing demands arising for reliable electronic equipment and systems' (Green, 1983, p. 3). One major tool of hazards analysis, the fault-tree analysis, was originally developed in the early 1960s for the aerospace industry: it was used in the Minute Man and Space and Missile Organization. Later it was used in Rasmussen's famous Reactor Safety Study (Green, p. 214).
9. See for example Ken Weller, *Lordstown and the Real Crisis in Production* (Solidarity, no date). Green-field plants are production plants newly established in areas – often rural – without a tradition of trade-union organization and militancy. John Mitchell of the Irish trade union IDATU has wryly observed how, in Ireland, multinational 'chemical plants are located in such trade union strongholds as Ballydine, Kilalla, Knockanish' (Jackson, *et al.*, 1985, p. 12). Bhopal may be considered a similar green-field site in India, lacking a tradition of trade-union organization.

10. This importance of workers' skills is confirmed by Buchanan and Bessant's report from a pigments manufacturing plant in Glasgow:

The operators in the 'automated' continuous process had thus not lost the responsibility they had in batch production. The ability to 'make or break' a batch previously lay with skilled operators who had a close relationship with the plant and often knew intuitively how to retrieve a batch. Production management were removed from this routine control and complained to research and development staff for not providing them with 'scientific' answers to reduce their dependence on local 'alchemy'. The first phthalocyanine plant had been kept in service because the old equipment and the experienced operators were able to produce blues although it was not known how the colour parameters were affected. In the oldest and dirtiest plant on the site there was a strong sense of pride among the operators who could make colour that could not be made elsewhere.

The sophistication of operator knowledge and intuition was indicated in the extremely complex routes that engineers had to use to replace them; for example, an automatic analyser used to determine a chemical coupling rate was based on a device used in kidney machines. (Buchanan and Bessant, 1985, p. 302)

11. Introduction of new technology is generally intended to redress the balance for management in this struggle. According to Buchanan and Bessant, 'recent research has indicated that some managers see in advanced computing technology new opportunities to reduce operator discretion and control'. Alas for the pious hopes of management, these researchers conclude that the introduction of this new technology does no such thing: 'The nature and limitations of the computer control system and the high cost of error made management *more* dependent on effective human intervention than with simpler batch production methods' (1985, p. 306).

12. *Each of the accidents we have reviewed occurred in the context of a process failure and whilst the men concerned were trying to maintain or restore production. In every case the dangerous situation was created in order to make it quicker and easier to do this. In every case the company's safety rules were broken. The process failures involved were not isolated events. Nor were the dangerous means used to deal with them. The men acted as they did in order to cope with the pressure from foremen and management to keep up production. This pressure was continual, process failures were fairly frequent and so the short-cutting methods used to deal with them were repeatedly employed. In each case it was only a matter of time before somebody's number came up. (Nichols and Armstrong, 1973, p. 20)*

Workers are therefore put in a situation by capital in which they are forced to evade safety rules. As Nichols and Armstrong note, these accidents are normal. In such situations, deaths become routine, part of the functioning of production, an inescapable result of the organization of work. In an examination of routine deaths in the British offshore oil industry, Wright (1986, p. 265) concludes that 'the hazardous situations in which the accident occurred were themselves largely the products of two aspects of the formal organization of work, the "speed-up" and the practice of "subcontracting".'

In some cases the safety rules are so unwieldy that workers conspire with managers to evade them. Management can draw up safety guidelines and correct procedures in the knowledge that workers will evade them to ensure production is uninterrupted. Thus, management has its cake and eats it: it gets uninterrupted production yet is covered in the case of accidents by its written safety procedures. If all the safety procedures are followed, production would grind to a halt. Infringements are inherent to production. Thus the Bhopal workers were told that if all the safety precautions were observed, production would be impossible.

Robert Jungk has illustrated this superbly in the case of the nuclear industry:

Patrice Fleury also confessed to me that the radiation safety men had long since grown tired of the unremitting routine of their jobs. There are too few of them to ensure that rules are obeyed, and if that were done consistently the whole operation would soon grind to a halt. '... Perfect safety exists only on paper. You hear about it only at scientific conferences and in the industry's sales literature,' Patrice Fleury said resignedly. 'Renouncing improvisation would mean even lower productivity.' (Jungk, 1979, pp. 9–10)

This is confirmed by the use of work-to-rule as a major weapon in industrial struggles.

13. *Capitalist wage work amounts to a complete surrender of one's creative powers when labour is purchased as a commodity for use in production. That period of forty hours a week is dead time; it belongs formally to capital, but the illusion remains that our off-hours are 'free-time' – free from capital's dictates. Rather, that period of our lives is where consumption of commodities and leisure takes place, simply mirroring wage work. With the advent of widespread drug testing, that relationship becomes even clearer; you smoke marijuana at a party on Saturday and you will pay the consequences on Monday morning after a piss test.* (The Fifth Estate, Summer 1986, pp. 17, 23)

14. For a basic critique of 'artificial intelligence' see Weizenbaum (1984).
15. In fact UCC is moving out of MIC altogether as part of its divestment policy. In July 1986 UCC announced its intention of selling off its agricultural products division.
16. Among other difficulties involved here is that old class struggle. 'First the mere mention of an interest in monitoring the performance of maintenance personnel might be sufficient to precipitate industrial strife' (Williams and Wiley, 1985, p. 356).
17. Elsewhere Shrader-Frachette (1985) has effectively demolished this position, which is a basic premise of most risk analysis and measurement.

18. See here Perrow, Noble and Jungk.

19. Pity that. The major advantages involved in having a chemical company executive living next door have been described in a modest proposal by Eve Epstein, staff writer with the *Charleston Gazette*:

> *The fastest way to prevent spills is to pass a law requiring all chemical-company executives and major stockholders to live next to their plants for at least ten months a year. I predict that after the first six months, the air in the Kanawha Valley will be almost as pristine as the air in – say – Danbury, Conn., home of Union Carbide's corporate headquarters.*
>
> An executive relocation plan would probably clear up communication problems, too. Last Sunday, residents at Institute complained they felt sick long before Carbide officials decided something was wrong. Think how much easier it would be to mosey on over to Warren Anderson's yard, lean across the fence and have a neighbourly chat with Carbide's chief executive officer.
>
> 'Yo, Warren. You having trouble breathing?'
>
> '(Cough, cough.)'
>
> 'The wife and kids, too?'
>
> '(Cough, cough.)'
>
> 'Then we're agreed that there's a bit of a leak here? Think we could get a few sirens going, maybe bring in a few oxygen masks? Thanks, buddy. Come over and borrow a couple cups of sugar anytime.' (Charleston Gazette, 17 August 1985, p. 5A)

20. See the recent spate of books on crisis and disaster management; for example, Kharbanda and Stallworthy, *Management Disasters and How to Prevent Them* (Gower, 1986), S. Fink, *Crisis Management: Planning for the Inevitable* (New York: American Management Association, 1986), J. M. Cronin, 'Plan for crisis communications', *Hydrocarbon Processing*, November 1985, pp. 226–30, and the two publications of Western Union, *When Every Second Counts ... Crisis Communications Planning* and *How to Develop a Crisis Communications Program*. See also 'Are you making the most of your crisis?' *Public Relations Journal*, June 1984, pp. 16–18, in which Dow Chemical Canada's director of corporate communications explains how to turn a crisis 'into a positive development that will enhance your credibility'.

Chapter 9

1. There has also been a drop in white-collar employment in the industry as well. Nearly 13,000 salaried workers, including chemical engineers and chemists, were laid off or retired early by the US chemical industry in 1985 alone. For more detailed figures by firm for the period 1980–5, see *Chemical and Engineering News*, 3 February 1986, p. 9.

Chapter 10

1. For Warren County, see B. Jacobs, 'The Carolina PCB protest', *East West Journal*, December 1982, pp. 36–41.

2. See for example, T. Fishlock, 'This has been our Hiroshima', *The Times* (London), 10 December 1984, p. 5, and Praful Bidwai, of the *Times of India*, who wrote, 'If there was a wretchedly undignified, hideously helpless form of megadeath after Hiroshima and Nagasaki, this is it.' *Newsweek* (17 December 1984, p. 12) reported, 'The next morning it looked like a neutron bomb had struck.' On the same theme, Lagadec noted of the chemical disaster at Flixborough in England, 'Some people, after having visited the devastated site, compared the disaster to what might have been caused by a minor atom bomb' (Lagadec, 1982, p. 22).

3. This similarity is not confined to MIC. See the press coverage of the use of nerve gas in the Iran–Iraq war, which stressed the similarities between organophosphorous pesticides and nerve gas and emphasizes the ease with which organophosphorous pesticide factories could be converted to produce nerve gases; for example, the *New York Times*, 30 March 1984, p. A7, and 31 March 1984, pp. 1, 4; the *Observer*, 11 March 1984, pp. 1, 11; *New Scientist*, 9 August 1984, p. 7; *Financial Times*, 8 August 1984, p. 2. This connection between nerve gases and pesticide production is historic (see for instance J. Borkin, *The Crime and Punishment of I.G. Farben*, Deutsch, 1979), as is the connection between chemical production and warfare (Davis, L.N., 1985, chapters 4 and 5). For a short useful list of such dual-purpose chemicals, with their civilian and military uses, see *South*, June 1986, p. 105.

4. Consider the composition of capital in Flixborough Nypro, owners of the Flixborough plant: Dutch State Mines (45 percent), British National Coal Board (45 percent), Fisons (10 percent).

5. A more extreme view has been formulated by *The Fifth Estate* in their description of the 'industrial plague'. See, 'PBB: case study of an industrial plague', *Fifth Estate*, May 1976; 'Capitalism's industrial plagues', *Fifth Estate*, September 1976; 'Mexican oil spill disaster: industrial plague widens', *Fifth Estate*, 22 October 1979; 'We all live in Bhopal', *Fifth Estate*, Winter 1985.

6. Thompson's formulation has been strongly criticized by, among others, *Midnight Notes*. See their 'Exterminism or class struggle?' in *No Clear Reason: Nuclear Power Politics*, Free Association, 1984, pp. 9–27.

7. Bahro himself did not feel the need to back up his position with extensive evidence. 'I don't intend to prove anything, to present the evidence for those who don't want to read the writing on the wall, as I believe that facts and arguments are not what such people lack' (Bahro 1982, p. 144).

8. A more up-to-date description of the situation was recently presented by Norman Edelman, a consultant with the American Lung Association:

> *According to Mr Edelman, each year 100,000 US workers die – and 340,000 are disabled – due to occupational diseases. In 1980, the Department of Labor estimated that job-related respiratory diseases alone accounted for at least 65,000 disabilities and 25,000 deaths.*
>
> *The likelihood of toxic exposure is high, he says. An estimated 1.2 million workers each year come into contact with potentially dangerous silica dust, 2.5 million workers who make or install asbestos products are at increased risk of developing lung cancer, and hundreds of thousands of textile workers exposed to cotton dust may contract a chronic and debilitating lung disease.* (Chemical Marketing Reporter, 30 March 1987, p. 21)

9. 'Bioavailability' is the degree to which a drug or other substance becomes available to the target tissue after administration.

10. For information on this see V. Hunt, *Work and the Health of Women*, Boca Raton, FA: CRC Press, 1979, chapter 6; *Pregnant Women at Work*, ed. G. Chamberlain, Royal Society of Medicine/Macmillan, 1984; Society of Occupational and Environmental Health, *Work and the Health of Women*, New York: SOEH, 1977; J. M. Stellman, *Women's Work, Women's Health: Myths and Realities*, New York: Pantheon, 1977, chapter 4; *Double Exposure: Women's Health Hazards on the Job and at Home*, New York: Monthly Review Press, 1984, pt 2; *Suffer the Children: Chemical Threats to Fertility*, Dublin: C. Crow, 1982. Bibliographies include National Institute for Occupational Safety and Health, *Comprehensive Bibliography on Pregnancy and Work*, US Department of Health, Education and Welfare, March 1978, and W. Chavkin and L. Welch, 'Occupational hazards to reproduction: an annotated bibliography', New York: Program in Occupational Health/Residency Program in Social Medicine, Montefiore Hospital, 1980. Good popular accounts which are not limited to the work-place include J. Elkington, *The Poisoned Womb*, Harmondsworth: Penguin, 1986 and C. Norwood, *At Highest Risk*, New York: Penguin, 1981. For a review of industrialization in relation to reproductive health see Jackson (1986).

11. See A. Roberts and Z. Medvedev, *Hazards of Nuclear Power*, Nottingham: Spokesman, 1977, pp. 21–2.

BIBLIOGRAPHY

There are four relevant bibliographies available: BARC, Mac Sheoin, Shrivastava and the Transnationals Information Centre, London. For the Bhopal killing itself, the best sources are Indian: Eklavya; *Bhopal – Industrial Genocide?;* the Centre for Science and the Environment report; and the continuing coverage in *Economic and Political Weekly* and in *Bhopal*, newsletter of the Bhopal Group for Information and Action. In the West, the best account is Everest; for continuing coverage, see *Chemical and Engineering News*, *Chemical Marketing Reporter* and the *New York Times*.

Castleman and Purkayastha have ably documented the double standards aspects, Stover is very useful on government and industry responses in the USA and Swift is the best source on the Design Institute for Emergency Relief Systems. Both APPEN and Pinto and O'Leary are collections of essential documents.

In typical neo-colonialist manner, this bibliography assumes all books are published in London unless otherwise stated.

Agarwal, A., et al. (1985) *No Place to Run*. New Market, TN/Delhi: Highlander Research Center/Society for Participatory Research in Asia.
Agnew, D. (1985) 'Bhopal: a case study in multinational mismanagement', *Safety and Risk Management*, May, pp. 10–21, 80.
Alliance for Safety and Health (1982) *Toxic Ireland*. Dublin: ASH.

Anon. (1976) 'How Union Carbide has cleaned up its image', *Business Week*, 2 August, p. 46.

—(1986) 'The cracks in Carbide's defence', *Business India*, 10–23 March, pp. 77–82.

APPEN (Asian Pacific Peoples' Environment Network) (1985) *The Bhopal Tragedy – One Year After*. Penang: Sahabat Alam Malaysia.

Bahro, R. (1982) *Socialism and Survival*. Heretic Books.

—(1986) *Building the Green Movement*. GMP.

Bang, R., and **Sadgopal, M.** (1985) 'Effects of the Bhopal Disaster on the Women's Health. An Epidemic of Gynaecological Disease'. P O Gopuri.

BARC (Bhopal Action Resource Centre) (1986) 'Bhopal and its aftermath: a selected bibliography'. New York: BARC, 10 July.

—(1986) 'Justice for the Bhopal victims: what next in the litigation and what can public interest/social action groups do?', Rev. ed. New York: BARC, 15 July. (Briefing Paper no. 2.)

Bellamy, L. (1985) 'How peoples' behaviour changes your plant operation', *Process Engineering*, July.

Bellamy, L. J. (1984) 'Not waving but drowning: problems of human communication in the design of safe systems', in *Ergonomics Problems in Process Operations*, Institution of Chemical Engineers Symposium Series no. 90, pp. 165–75.

Bhopal – Industrial Genocide? (1985) Hong Kong: Arena.

Bhopal. Newsletter of the Bhopal Group for Information and Action.

Bidwai, P. (1986) Interview with Les Levidow, unpubl.

Brouwers, A.A.F. (1984) 'Automation and the human operators: effects in process operations', in *Ergonomics Problems in Process Operations*, Institution of Chemical Engineers, Symposium Series no. 90, pp. 179–90.

Buchanan, D.A., and **Bessant, J.** (1985) 'Failure, uncertainty and control: the role of operators in a computer integrated production system', *Journal of Management Studies* 22, 3: 292–308.

Cardan, P. (1969) *The Meaning of Socialism*. Solidarity.

Castleman, B. I., and **Purkayastha, P.** (1985) 'The Bhopal disaster as a case study in double standards', in J. Ives, ed. *The Export of Hazard*. Boston: Routledge & Kegan Paul, pp. 213–22.

Centre for Science and the Environment (1985) *Report on the Indian Environment 1984–1985*. New Delhi: CSE.

Commoner, B. (1979) *The Closing Circle: Nature, Man and Technology*. New York: Bantam.

Craig, A. (1984) 'Process control operations as a watchkeeping task', in *Ergonomics*

Problems in Process Operations, Institution of Chemical Engineers Symposium Series no. 90, pp. 36–7.

Dakin, S. (1985) 'Bhopal: a personal view', in V. Pinto and S. D. O'Leary, eds *Reprints of Selected Documents on the December 1984 Union Carbide Chemical Incident*. San Francisco: Washington Research Institute.

Davenport, J.A. (1977) 'Survey of vapour cloud incidents', *Chemical Engineering Progress*, September, pp. 54–63.

Davis, D.L., *et al.* (1981) 'Estimating cancer causes: problems in methodology, production and trends', in *Quantification of Occupational Cancer*. Cold Spring Harbor, NY: Cold Spring Harbor Laboratory.

Davis, L.N. (1984) *The Corporate Alchemists: Power and Problems of the Chemical Industry*. Martin Robinson.

Davis, M. (1982) 'The impact of workplace health and safety on black workers', in F. Goldsmith and L. E. Kerr, *Occupational Safety and Health*. New York: Human Sciences.

Delhi Committee on Bhopal Gas Tragedy (1985) *Repression and Apathy in Bhopal*. Delhi: DCBGT.

Delhi Science Forum (1985) *Bhopal Gas Tragedy*. Delhi: DSF.

Dembo, D., and **Morehouse, W.** (1986) 'Judge Keenan's decision on venue in the Bhopal case'. New York: BARC, 23 May.

Eberts, R.E. (1985) 'Cognitive skills and process control', *Chemical Engineering Progress*, December, pp. 30–4.

Eklavya (1984) *Bhopal – A People's View of Death, Their Right to Know and Live*. Bhopal: Eklavya.

Embrey, D. E. (1981) 'Approaches to the evaluation and reduction of human errors in the process industries', in *Reliable Production in the Process Industries*, Institution of Chemical Engineers Symposium Series no. 66, pp. 181–94.

Engler, R. (1985) 'Many Bhopals: technology out of control', *Nation*, 27 April, pp. 488–500.

Epstein, S. S. (1978) *The Politics of Cancer*. San Francisco: Sierra Club.

Everest, L. (1985) *Behind the Poison Cloud: Union Carbide's Bhopal Killing*. Chicago: Banner.

Faulkner, A. J. (1984) 'The operator in a highly-automated plant', in *Ergonomics Problems in Process Operations*, Institution of Chemical Engineers Symposium Series no. 90, p. 235.

Fishbein, L. (1981) 'Overview of some aspects of quantitative risk assessment', in *Occupational Cancer and Carcinogenesis*. Washington, DC: Hemisphere, pp. 355–86.

Flores, A. (1983) 'Safety in design: an ethical viewpoint', *Chemical Engineering Progress*, November, pp. 11–14.

Frankel, M. (1981) *A Word of Warning – The Quality of Chemical Suppliers' Health and Safety Information*. Social Audit.

Fulano, T. (1981) 'Against the megamachine', *The Fifth Estate*, July, pp. 4–5.

George, L. N. (1982) 'Love Canal and the politics of corporate terrorism', *Socialist Review* 12, 6: 9–38.

George, S. (1976) *How the Other Half Dies*. Harmondsworth: Penguin.

Gersuny, S. (1981) *Work Hazards and Industrial Conflict*. Hanover, NH: University Press of New England.

Green, A. H. (1983) *Safety System Reliability*. Chichester: Wiley.

Haastrup, P. (1983) 'Design error in the chemical industry. Case studies', in *Loss Prevention and Safety Promotion in the Process Industries*, vol. 1, *Safety in Operations and Processes*, Institution of Chemical Engineers Symposium Series no. 80, pp. J15–27.

Halle, D. (1984) *America's Working Man: Work, Home and Politics among Blue-collar Property Owners*. University of Chicago Press.

Hanson, J. L. (1977) *A Dictionary of Economics and Commerce*, 5th ed. MacDonald & Evans.

Hayashi, Y. (1985) 'Hazard analysis in chemical complexes in Japan – especially those caused by human error', *Ergonomics* 28, 6: 825–41.

Holmes, K. (1982) 'The export of hazardous products and industries: a bibliography', *Int. J. Health Services* 12, 3: 517–21.

Howard, H. B. (1983) 'Efficient time use to achieve safety of processes or "How many angels can stand on the head of a pin?" ', in *Loss Prevention and Safety Promotion in the Process Industries*, vol. 1, *Safety in Operations and Processes*, Institution of Chemical Engineers Symposium Series, no. 80, pp. A11–19.

Human Reliability Assessment Group (1985) 'Closing the gap between ergonomics and engineering', *Process Engineering*, July.

ICFTU/ICEF (1985) *The Trade Union Report on Bhopal*. Brussels/Geneva: ICFTU/ICEF.

Jackson, P. (1986) *Industrialization and Reproductive Rights*. Florence: European University Institute.

Jackson, P., et al. (1985) *No More Bhopals*. Dublin: Alliance for Safety and Health.

Jones, T. (1986) 'Hazards for export: double standards?', in L. Levidow, ed. *Science as Politics*. Free Association, pp. 119–38.

— (1987) 'Bhopal: backward or advanced?' in L. Levidow and D. Gill, eds. *Anti-racist Science Teaching*. Free Association, pp. 283–92.

Jungk, R. (1979) *The Nuclear State*. Calder.

Kinnersly, P. (1973) *The Hazards of Work: How to Fight Them*. Pluto.

Kletz, T. (1985a) *What Went Wrong? Case Histories of Process Plant Disasters*. Houston, TX: Gulf Publishing, pp. 1091–3.

— (1985b) 'Making a chemical plant safer', *Chemistry in Britain*, December.

Lagadec, P. (1982) *Major Technological Risk*. Oxford: Pergamon.

Lamb, J. (1986) 'Scientists rush in where computers fear to tread', *Observer*, 19 October, p. 43.

Levidow, L. (1979) 'Three Mile Island: the ideology of "safe level" as a material force', *Radical Science Journal* 9: 82–92.

MacKenzie, D. (1986) 'Chemical disasters need not happen', *New Scientist*, 5 June, pp. 42–5.

Mac Sheoin, T. (1985) 'The export of hazardous products and industries: a bibliography', *Int. J. Health Services* 15, 1: 145–55.

— (1986) 'Bhopal: a bibliography', *Int. J. Health Services* 16, 3: 441–68.

— (1987) 'The export of hazardous products and industries: a bibliography', *Int. J. Health Services* 17, 2: 343–64.

Medico-Friend Circle (1985) *The Bhopal Disaster Aftermath: An Epidemiological and Socio-medical Survey*. Bangalore: MFC.

Menzies, H. D. (1978) 'Union Carbide raises its voice', *Fortune*, 25 September, pp. 86–9.

Midnight Notes (1980) 'The work/energy crisis and the apocalypse', *Midnight Notes* 2, 1.

Morehouse, W., and **Subramaniam, A.** (1985) *The Bhopal Tragedy*. New York: Council on International and Public Affairs.

Narayan, T., et al. (1985) 'Rationale for the use of sodium thiosulphate as an antidote in the treatment of the victims of the Bhopal gas disaster – a review'. Bhopal: Medico-Friend Circle, 7 June.

Nelson, W. R., and **Jenkins, J. P.** (1985) 'Expert system for operator problem solving in process control', *Chemical Engineering Progress*, December, pp. 25–9.

Nichols, T. and **Armstrong, P.** (1973) *Industrial Accidents and the Conventional Wisdom*. Bristol: Falling Wall.

Noble, D. F. (1980) 'Cost-benefit analysis', *Health/PAC Bulletin*, July, pp. 1–2, 7–12, 27–40.

Nostrom II, G. P. (1982) 'Fire/explosion losses in the CPI', *Chemical Engineering Progress*, August, pp. 80–7.

Oil, Chemical and Atomic Workers' International Union (1986) 'Background' Denver, CO: OCAWIU, 4 April.

Organization for Economic Co-operation and Development (1983) *Economic Aspects of International Chemicals Controls*. Paris: OECD.
Perrow, C. (1982) 'The President's Commission and the normal accident', in *Accident at Three Mile Island: The Human Dimensions*. Boulder, CO: Westview, pp. 173–84.
— (1984) *Normal Accidents: Living with High-risk Technology*. Basic.
Pinto, V., and **O'Leary, S. D.**, eds (1985) *Reprints of Selected Documents on the December 1984 Union Carbide Chemical Incident*. San Francisco: Washington Research Institute.
Pomata, G. (1979) 'Seveso: safety in numbers?', *Radical Science Journal* 9: 69–81.
Prakash, P. (1985) 'Failure of scientific community', *Economic and Political Weekly*, 25 May, pp. 909–10.
Rasmussen, J. (1982) 'Human reliability in risk analysis', in A. E. Green, ed. *High Risk Safety Technology*. Chichester: Wiley, pp. 143–70.
— (1985) 'Trends in human reliability assessment', *Ergonomics* 28, 8: 1185–95.
Rausch, D. A. (1986) 'Operating discipline: a key to safety', *Chemical Engineering Progress*, June, pp. 13–15.
Reiko, W. (1985) 'Autopsy of a killing: the Bhopal incident', *AMPO* 17, 2: 56–61.
Sass, R., and **Crook, G.** (1981) 'Accident proneness: science or non-science?', *Int. J. Health Services* 11, 2: 175–90.
Sella, Jr, G. J. (1986) 'Strengthening CMA advocacy'. Washington, DC: Chemical Manufacturers' Association.
Shepherd, A. J. (1982) 'Task analysis, training development and some implications for process plant design', in *Design 82*, Institution of Chemical Engineers Symposium Series no. 76, pp. 151–60.
Shrader-Frachette, K. (1985) 'Technological risk and small probabilities', *J. Business Ethics* 4: 431–45.
Shrivastava, P. (1986) 'Bibliography of publications on Bhopal'. New York: Industrial Crisis Institute, December.
— (1987) *Bhopal: Anatomy of a Crisis*. Cambridge, MA: Ballinger.
Slater, D. M. (1978) 'Vapour clouds', *Chemistry and Industry*, 6 May.
Stover, W. A. (1985) 'A field day for the legislators', paper presented to the Chemical Industry After Bhopal Conference, London, November. Oyez International.
Suser, I. (1985) 'Union Carbide and the community surrounding it: the case of a community in Puerto Rico', *Int. J. Health Services* 15, 4: 561–83.
Susman, P., *et al.* (1983) 'Global disasters, a radical interpretation', in K. Hewitt, ed. *Interpretations of Calamity*. Boston: Allen & Unwin.
Swift, I. (1985) 'The engineering approach to safe plants', paper presented to the

Chemical Industry After Bhopal Conference, London, November. Oyez International.

Talty, J. T. (1986) 'Integrating safety and health issues into engineering school curricula', *Chemical Engineering Progress*, October, pp. 13–16.

Teague, H. J. (1985) 'Cyanide poisoning of Bhopal victims – a hypothesis', unpublished paper available from Harold J. Teague, Pembroke State University, Pembroke, North Carolina 28372, USA.

Thomas, L. M. (1986) *Why We Must Talk about Risk: A Personal View*. Washington, DC: Environmental Protection Agency.

Transnationals Information Centre London (1986) 'Bhopal – a selected bibliography'. TIC.

Union Carbide Corporation (1985a) 'Bhopal: crisis management and communications at Union Carbide Corporation. Danbury, CN: UCC.

— (1985b) 'Bhopal methyl isocyanate incident investigation team report'. Danbury, CN: UCC.

Vanaik, A. (1986) 'The Indian left', *New Left Review*, September/October, pp. 49–70.

Van Mynen, R. (1985) 'Van Mynen presentation – Bhopal MIC incident', March 1985 press conference, 20 March. Danbury, CN: Union Carbide Corporation.

Varadarajan, S., *et al.* (1985) 'Report on scientific studies on the factors related to Bhopal toxic gas leakage'. December.

Wagoner, J. K. (1976) 'Occupational carcinogenesis: the two hundred years since Percivall Potts', *Annals of the New York Academy of Sciences* 271: 1–4.

Wartenburg, D. (1985) 'Review of Ives, J. (ed.), "The export of hazard" ', *Science for the People*, November, pp. 30–1.

Watson, I. A. (1985) 'Review of human factors in reliability and risk assessment', in *Assessment and Control of Major Hazards*, Institution of Chemical Engineers Symposium Series no. 93, pp. 323–51.

Weizenbaum, J. (1984) *Computer Power and Human Reason*. Harmondsworth: Penguin.

Williams, L. J. C., and **Wiley, J.** (1985) 'Quantification of human error in maintenance for process plant probabilistic risk assessment', in *Assessment and Control of Major Hazards*, Institution of Chemical Engineers Symposium Series no. 93, pp. 353–65.

Wills, G. A. (1973) *A Guide to Process Engineering with Economic Objective*. Aylesbury, Bucks: Leonard Hill.

Wilson, D. C. (1980) 'Flixborough versus Seveso – comparing the hazards', in *Chemical Process Hazards with Special Reference to Plant Design*, Institution of Chemical Engineers Symposium Series no. 58, pp. 193–207.

Witt, M. (1979) 'Dangerous substances and the US worker: current practice and viewpoints', *Int. Labour Review*. March/April, pp. 165–77.

Wolfe, S. M. (1977) 'Standards for carcinogens: science affronted by politics', in *Origins of Human Cancer*. Cold Spring Harbor, NY: Cold Spring Harbor Laboratory, pp. 1735–47.

Wood, K. (1985) 'Avoiding engineering disasters', *Health & Safety at Work*, May, pp. 20–2.

Woods, P. D. (1984) 'Some results on operator performance in emergency events', in *Ergonomics Problems in Process Operations*, Institution of Chemical Engineers, pp. 21–32.

Wright, C. (1986) 'Routine deaths: fatal accidents in the oil industry', *Sociological Review*, pp. 265–89.

Wyrick, B. (1981) 'How job conditions led to a worker's death', in *Hazards for Export*. Long Island, NY: Newsday, pp. 23–25R.

INDEX

acetone 204
acitol 148
Acmat Corp. (USA) 200
acrylonitrile 156, 286
Adler, S. J. 124, 141
A. D. Little (USA) 175
Aerflex Corp. (USA) 307n.
aerosols 288
AFL-CIO (USA) 155, 188
Agape (India) 304n.
Agent Orange 284–5, 304n.
Agnew, D. 27, 28, 42, 228
Agricultural Ministry (India) 59
Ahmad, N. K. 50
A. H. Robins (USA) 131, 133, 210–12
Aitala, R. 44
Alarie, Y. C. 119
aldicarb 171, 205
aldicarb oxime 166, 168–71, 173, 180–1, 205, 306n.
Alliance for Safety and Health 8
Allied Chemical (USA) 212
All-Indonesian Labour Federation 40
Allstate Insurance (USA) 203

alpha naphthol 68
aluminium 288
American Casualty Excess Insurance Co. 200–1
American Chemical Society 119
American Civil Liberties Union 234
American Conference of Government and Industrial Hygienists 25
American Cyanimid 38, 149, 228
American Institute of Chemical Engineers 162, 182–3, 211
American International Group 200
American Lawyer 45, 85, 129
American Motors 201
Ames test 25
ammonia 83, 147, 181, 211, 269, 287
anaesthetic gases 288
Anderson, W. 14–15, 18, 30–1, 36–7, 43, 45–6, 126–7, 132, 135–6, 138–9, 152, 155, 163–4, 166, 171–2, 198, 250, 298n., 301n., 304n.
angiosarcoma 278
Ankleswaria, S. 96
antioxidants 286

326 CORPORATE KILLING

Antwerp (Belgium) 151, 184, 204
APPEN 76, 101, 124
Arena Press (Hong Kong) 76
Arpège 171
arsenic 278, 288
arsine 248
arson 262
asbestos 125, 130, 134, 177, 199, 200, 203, 278–9, 304n., 315n.
asbestosis 278
Atrazine 265–6
Avashia, B. 52, 109, 111–12, 169
Ayyager, N. A. 60

Bahro, R. 271, 277, 280, 282
Bailey, F. L. 132
Ballantyne, B. 27, 111, 273, 297–8n.
Ballydine, Co. Tipperary (Ireland) 218, 306n.
Banerji, D. 61
Bang, R. 100, 119
Bannerjee, S. 81
BASF (West Germany) 252–8, 265, 268–9
Basle (Switzerland) 1, 4, 201, 259, 260, 264
Baton Rouge, Louisiana (USA) 156, 197
Bayer (West Germany) 28, 151, 265–6, 269
Baygon 108
BBC (Britain) 70
Bellamy, L. 237
Bellamy, L. J. 240
Belle, West Virginia (USA) 196
Beller, G. E. 165
Belli, M. 122, 128
benzene 156, 278, 286–7
Bergman, D. 87, 303n.
Béziers (France) 2, 150, 299n.
Bhandari, Dr 61, 109, 111
Bharat Heavy Electricals Ltd (India) 100
Bhopal Disaster Monitoring Group (Japan) 154
Bhopal Gas Leak Disaster (Processing of Claims) Act, 1985 123, 125
Bhopal Group for Information and Action 85, 92, 303–4n.

Bickett, P. B. 198
Bidwai, P. 57, 71, 73, 91, 105
Biesterfeld US Inc. (West Germany) 200
Bishop, D. R. 153
Black June Movement (India) 46, 301n.
Bombay 2, 147
Borg Warner (USA) 156
Bose, T. 75, 89
Brandt, W. 259
Brandweir, D. 211
British Medical Research Council Perceptual and Cognitive Performance Unit 229
British Nuclear Fuels Ltd 298n.
British Safety Council 28
Brock, L. 199
Brock, W. 176–7
Browning, J. 17, 19, 27, 33–4, 36, 46, 48, 170, 271
Brown, W. 22, 107
Bucher, J. 107
Burston Marstellar 32
Business Europe 260
Business India 77, 110–11, 300n.
Business Week 16, 30, 32, 44, 97–8, 124, 134–5, 152, 245, 262
butadiene 156, 287
Byrd, Senator 167

cadmium 268–9
CAER/Community Awareness and Emergency Response programme (USA) 158, 161, 193
Calico Mills (India) 148
carbaryl 43
carbon dioxide 50, 83
carbon monoxide 50–1, 112
carbon tetrachloride 68
Cardan, P. 233
Caritas 56
Carling (France) 178
Carnegie Mellon University (USA) 22, 107
Carnsore Point (Ireland) 285
Carol, M. H. 119

Carrh, B. W. 188
Castleman, B. 77, 274
caustic soda 66, 302n.
Cdf. Chémie (France) 178
Center for Chemical Plant Safety (USA) 162
Centers for Disease Control (USA) 23–4, 90, 92, 109, 170, 205, 296n.
Central Bureau of Investigation (India) 26, 44, 62, 66, 69, 70, 82, 109, 111, 117, 300n.
Central Electricity Generating Board (Britain) 213
Central Railway Hospital (Bhopal) 305
Centre for Science and the Environment (New Delhi) 60, 149
Centre for Social Medicine (New Delhi) 296n.
Chandler, E. 164
Chandra, H. 21, 50, 84, 109–10
Channel 4 (Britain) 29
Chapter 11 (US bankruptcy code) 16, 131, 133, 201–2
Charleston Gazette 170–1, 173
Charleston, West Virginia (USA) 177, 187
Chembur, Bombay 148
Chemical and Engineering News 26–7, 38, 82, 112, 119, 137–8, 139, 234
Chemical Co. of Malaysia 300n.
Chemical Engineering Progress 182
Chemical Industries Association Ltd (Britain) 1, 25, 153
Chemical Industry Council (USA) 159
Chemical Industry Institute of Toxicology (USA) 14, 25–6
Chemical Manufacturers' Association (USA) 4, 30, 145, 152, 157–9, 161, 170, 181, 191–2, 195–6, 212
Chemical Marketing Reporter 53
Chemical Week 124
Chernobyl 212–13, 238, 249, 268, 305n.
Chesley, S. 132
Chile 150
Chitre, A. 63
chloralkali industry 152

chloride 265
chlorinated hydrocarbons 265, 268
chlorine 68–9, 147, 180, 204, 255
chlorobenzene 268
chloroform 68, 83, 107
chlorosulphuric acid 69, 204
Christopherson, P. 133
CIA 81
Ciba-Geigy (Switzerland) 153, 262, 264–6, 297n.
Cimanggis (Indonesia) 39–40
Ciresi, M. 136
CITGO Petroleum Corp. (USA) 193, 196
Citizens' Commission on Bhopal (USA) 77, 140
Clay Cross Co. (Britain) 297n.
Clean Sites Inc. (USA) 157, 169, 306n.
coal dust 278
Coale, J. 122, 128, 132
cobalt 231
Cochin (India) 148
Colbert, H. I. 17, 52–3
Commission of Inquiry (Madhya Pradesh) 69, 82
Commoner, B. 280
Communist Party – Marxist (India) 74
Confederation of British Industry 234
Congress (I) Party (India) 62, 73, 80–1
Cooley, M. 227
copper 288
Copper-7 202
Corbett, H. J. 151
Corporate Accountability Research Group (USA) 275
cortico-steroids 113
Council for Scientific and Industrial Research (India) 35, 51, 59, 63, 67, 147, 309n.
Council on Environmental Quality (USA) 144
Cox, G. 157, 170, 181
Craig, A. 229
Crum & Foster Insurances Cos (USA) 198
Cubatão (Brazil) 1, 141, 149, 211, 271, 302n.

cyanide 15, 50, 64, 83, 90, 94–5, 106–7, 109, 111–14, 287, 307n.
cyanogen pool 94–5, 103

Dalkon Shield 130–1, 133, 201–2
Danbury, Connecticut (USA) 30, 37, 52
Das, S. K. 84
Dass, I. 27–8, 89, 94, 117
Daunderer, M. 109–10
Davenport, J. A. 209
Davies, S. C. 261
Davis, D. L. 279
Davis, L. N. 226
Deepwater, New Jersey (USA) 225
Defence Research and Development Establishment (India) 298n.
Delhi Committee on the Bhopal Gas Tragedy 76, 302n.
Delhi Science Forum 24, 26–7, 29, 58, 61–2, 74–5, 77
Dembo, D. 137
Department of the Environment (India) 31
Department of Science and Technology (India) 29
Deputy Inspector-General Hospital (Bhopal) 104, 115
Design Institute on Emergency Relief Systems 182–3, 185, 207, 307n.
Devnia (Bulgaria) 264
dextran 286
Dhanda, R. P. 60
Diamond Shamrock (USA) 196
Digital Equipment Corp. (USA) 280
dioxin 204, 259
disinfectants 288
Donovan Leisure Newton & Irvine 127
Dormagen (West Germany) 151
Dow Chemical (USA) 9, 156, 184, 187, 212, 220
Dow Laboratories 297n.
Doyer, D. 187
Draper, E. 191
Drug Action Forum 75
Dunbar, West Virginia (USA) 168

Dunphy, L. 169
Du Pont (USA) 25, 36, 38, 151, 185, 188, 196, 200, 217, 225, 234–5
Dutch State Mines 314n.

Eberts, R. E. 228
Economic and Political Weekly 89, 105
Edelman, N. 315n.
Efstathiou, J. 238
E, G & G (Idaho) Inc. (USA) 239
Eklavya (Bhopal) 60, 74
Elanco 149
electronics industry 247, 279–80
Eli Lilly 149
Elizabeth, New Jersey (USA) 229
Ellery, Y. 24
Embrey, D. E. 215, 224, 238
emergency relief systems 182–6, 307n.
Environmental Action 133
Environmental Mediation International 122
Environmental Pollution Control Board (Madhya Pradesh) 59
Environmental Protection Agency (USA) 26, 155, 157–8, 169, 175, 177, 188, 191, 197, 203–4, 209, 245, 305n., 308n.
Epps, T. 165–6, 180
Epstein, E. 313n.
Epstein, S. 277
ethyl acetate 203
ethylene 203
ethylene glycol 265
ethylene oxide 205, 255, 307n.
European Chemical News 196
European Council of Chemical Manufacturers' Associations 153–4, 201
European Environment Bureau 210
Everest, L. 64
Eye Hospital (Bhopal) 305n.
Eyer, J. 280

Fairfax, Virginia (USA) 187
Falk, H. 296n.

INDEX   329

Faulkner, J. 229
Fazel, M. 252
Federation of India Chambers of
   Commerce and Industry 56
Feldman, M. B. 127
Fernandez, L. 191
Film Recovery Systems (USA) 114
Financial Times 12, 20, 52, 131, 201-2
Fisher, H. 182
Fisons 314n.
flame retardants 287
Flamm, A. 15
Flixborough, Humberside (Britain) 1, 153,
   213, 217, 309n., 314n.
Florio, J. 156
FMC (USA) 36, 151
Food and Drug Administration (USA)
   308n.
Ford (USA) 200-1
Foreign Exchange Regulations Act (India)
   38, 145

GAF Corp. (USA) 131, 133-5, 139, 163
Gagliardi, D. V. 188
Galanter, M. 124, 138
Gandhi, R. 5, 54, 57, 62, 73, 80, 91, 94,
   115-16, 123, 302n.
Gandhi Medical College (Bhopal) 51, 61,
   93, 96, 103, 110, 113, 124
Gauley Bridge, West Virginia (USA) 299n.
Gaydos, J. 204
Geismar, Louisiana (USA) 247, 252-3,
   255-6
Gelinas, G. 180
General Electric (USA) 123, 200
General, Municipal, Boilermakers' and
   Allied Trades Union (Britain) 153, 282
Gibson, J. E. 26
Givaudin 303n.
Gladwin, T. W. 37, 39
Glow Plastic Corp. (USA) 307n.
Gokhale, V. 48, 125
Gore, Oklahoma (USA) 1, 212
Gothoskar, S. 275

Green, G. 23, 90-1, 113
Green Party (West Germany) 258, 261,
   268, 282
Greenpeace 213, 260, 266
Gremillion, E. F. 255
Gresham, J. 171
Grover, A. 30, 50, 114, 147
Guardian 276
Gujarat (India) 146
Gujarat Pollution Control Board 146
Gupta, V. K. 45
Gwalior (India) 27, 148

Haastrup, P. 219-21
Haayen, R. 203
Hager, R. 122, 128, 131, 133
hair dyes 288
Halberg, G. P. 22
Halle, D. 224, 226, 230, 232
Hamidia Hospital (Bhopal) 21, 50, 55, 62,
   105, 110, 112
Hanrahan, J. 218, 300n.
Happel, J. 222-3
Harrisburg, Pennsylvania (USA), *see also*
   Three Mile Island 4, 68
Hasan, A. 40
Health and Safety Executive (Britain) 153,
   283, 297n.
Henderson, D. 169
Henderson, Kentucky (USA) 171
Henderson, T. W. 134
Henry, J. 25
Hercules (USA) 307n.
Heyman, S. 134-5
Heyndrickx, Prof. 24
Highlander Research Center (USA) 77
Hill & Knowlton (USA) 256
Hilton, R. 30
Hindu 54
Hindustan Times 68
Hiroshima 4, 273, 314n.
Hitavada 115
Hoechst (West Germany) 265, 268-9
Hoffinger, J. 121

Hoffman LaRoche (Switzerland) 185, 262, 265, 303n.
Holman, B. G. 53, 123, 125–6, 129, 132, 164, 301n.
Home Office (India) 21
Houk, V. 170
Howard, E. 173, 216–17
Huningue (France) 264
Hurni, B. 266
hydrochloric acid 255
hydrocyanic acid 51, 287
hydrofluoric acid 250, 307n.
hydrogen chloride 307n.
hydrogen cyanide 27, 50–2, 83, 94, 112, 298n.

IBM (USA) 200
ICI (Britain) 229, 240, 300n.
India Cloth Merchants' Association 56
Indian Chemical Manufacturers' Association 53
Indian Council on Medical Research 51, 61, 84, 93–6, 101–6, 110–11, 113–15, 118–19, 298n.
Indian Express 41, 104, 112
Indian Journal of Medical Research 105
Indian Medical Association 115
Indian Toxicology Research Centre 107, 119
Indo-American Chamber of Commerce 146–7
Industrial Risk Insurers 211
Industrial Toxicology Research Centre (Lucknow, India) 56
Institute, West Virginia (USA) 1–2, 4, 17–18, 33, 35–6, 39, 43–4, 51, 77, 156–7, 164–81, 187, 189, 192, 196, 203–5, 222, 242, 247, 271, 273–4, 299n., 302n., 306n.
Institute of International Studies (USA) 30
International Association of Machinists and Aerospace Workers 171, 176
International Brotherhood of Painters and Allied Trades 161

International Chemical Workers' Union 253
International Confederation of Free Trade Unions 281
International Finance Corp. (World Bank) 249
International Labour Organization 281
Ixhuatapec (Mexico) 1, 55, 141, 211, 271, 309n.

Jabalpur (India) 148
Jaeger, C. 23
Jaiprakash Nagar (Bhopal) 55
Jana Swaasthya Kendra/People's Health Clinic (Bhopal) 75, 79, 80, 115–17
Janata Party (India) 31
Jenkins, J. O. 239
Jeuck, J. 30
Johns, A. 23
Johns Manville (USA) 130, 199
Johnson & Higgins (USA) 198
Johnson & Johnson (USA) 32
Joliet, Illinois (USA) 193
Jones, J. M. 212
Jordan, D. G. 222–3
Jourdain, L. 154
Jungk, R. 312n.

Kalyan (India) 148
Kanawha Valley, West Virginia (USA) 156, 165, 172, 197
Kanawha Valley Emergency Planning Council (USA) 194
Kanpur (India) 148
Kaplan, J. 23–4, 92
Karawan, H. 194
Karnataka (India) 146
Keenan, Judge 42, 126–8, 130, 132–4, 136–7
Kelley, Drye & Warren (USA) 304n.
Kennedy, R. 170, 173–4
Kerala (India) 146–7

INDEX 331

Kerala Sastra Salutya Parishad (India) 296n.
Kerr McGee (USA) 212
Khan, A. 296
Khan, N. M. 124
Kharabandan, O. 188
Killala, Co. Mayo (Ireland) 2
King Edward Memorial Hospital (India) 102, 104
Klatte, E. 210
Kletz, T. 209, 240, 242
Krishnamoorthy, B. R. D. 42
Krishna Murthi, C. R. 107
Kumaraswami, S. 35
Kurchatov Atomic Energy Institute (USSR) 212

Labour Commissioner (New Delhi) 147
Lagadec, P. 202, 215, 309n.
Lake Charles, Louisiana (USA) 193, 196
Lal, Dr 42
La Littorale (France) 150
Lancet 26
Lang, R. 144
Lautenberg, Senator 188
Lawyers' Collective (Bombay) 50, 75, 114, 116, 129, 147
lead 288
Legasov, V. 212
Leone, F. 168
leukaemia 278
Leverkusen (West Germany) 269
Linden, New Jersey (USA) 156, 211
liquefied natural gas 181
liquefied petroleum gas 211
Livingston New Town (Scotland) 247-8, 272, 299n.
Local Intelligence Bureau (Bhopal) 81
Lok Sabha (India) 73
Lonza 265
Love Canal, New York (USA) 1, 4, 57, 65, 192, 271
Lowy, A. 132
Luxton, S. 108

MacKenzie, D. 223
Madhya Pradesh 67, 69, 78-9, 81-2, 84, 87, 96, 104, 110, 115-17, 122, 146, 148
Madhya Pradesh Health Department 93, 111
Madurai Coats (India) 148
Maharashtra Pollution Control Board 146
malathion 228
Malle (Belgium) 204
Manhattan Project 299n.
Marshall Teichner 124
Marsh & McLennan 199, 200, 209
Mathur, Dr 94
Mazzocchi, T. 280
Medical Research Council Toxicology Unit (Britain) 106
Medico-Friend Circle 74-5, 91, 93, 95-6, 99-101, 103-4, 106
Medico-Legal Institute of Gandhi Medical College (Bhopal) 113
Melius, J. 92
Merck, Sharp & Dohme (Ireland) 150, 218, 300n., 306n.
mercury 40, 152, 261, 269
Merriman, P. 25
mesityl oxide 204
mesothelioma 278
Metamoros (Mexico) 211
methanol alcohol 265
methylamine 118
methylene chloride 156, 306n.
methyl vinyl ketone 265
Metzenbaum, Senator 161
Mexico 18
Mexico City 141
Michigan State (USA) 1
MIC (methyl isocyanate)
  leaks 18, 45ff., 59, 83-4, 164-5, 204, 306n.
  storage/production problems 9-12, 35-6, 42ff., 63-8, 139, 149-56, 166-9, 175-81, 240, 284, 297n., 312n.
  toxicity 20-8, 56ff., 89-120, 170-3, 273, 296n., 304-5n.

Midland, Michigan (USA) 156
Miller, D. 170
Miller, R. 255
Minamata (Japan) 1
Ministry of Chemicals and Fertilizers (India) 66
Ministry of Defence (Britain) 27
Ministry of Labour (Japan) 154
Mirer, F. E. 155
Misra, Dr 51, 94, 96–8, 103–4, 110–11, 113, 124
Mitsubishi (Japan) 28, 152
Mitsui Toatsu Chemicals (Japan) 154
Mohan, Prof. 60
Mokhiber, R. 275
Molinari, G. 156
monomethylamine 68, 109
Monsanto (USA) 151, 153, 157–8, 167, 191–2, 196, 299n.
Morehouse, W. 77, 137
Moret, M. 268
Morton Thiokol (USA) 151
Mukund, J. 20, 39, 296n.
Multinational Monitor 128, 131
Munde, W. 268
Munich Institute of Toxicology (West Germany) 109
Munich Reinsurance Co. (West Germany) 202
Munoz, E. 35
Muttenz, Basle (Switzerland) 259
MYCIN 239

Nagasaki 4, 273, 314n.
Nagrik Rahat aur Punarwas Committee (India) 70, 75, 80, 82, 84, 89, 115, 117
Nagu, M. 110–11, 115, 304n.
Nai Duniya 111
National Bureau of Plant Genetic Resources (India) 29
National Campaign Against Toxic Hazards (USA) 191
National Campaign Committee on Bhopal (India), *see* Rashtria Abhiyan Samiti

National Chemical Laboratory (India) 60
National Coal Board (Britain) 314n.
National Institute for Chemical Studies (USA) 165–6, 172
National Institute of Environmental Health Science (USA) 107, 119
National Institute of Occupational Safety and Health (USA) 92, 221, 279
National Labour Relations Board (USA) 253
National Poisons Information Centre (Britain) 25
National Resources Defense Council (USA) 50, 187, 189
National Toxicology Programme (USA) 25–6
National Wildlife Federation (USA) 157
natural gas 307n.
Nayak, G. C. 53
Nelson, W. R. 239
Nemery, B. 106–7
New Jersey Environmental Protection Agency (USA) 160
New Jersey (USA) 156, 159–60, 197, 228
Newsweek 149
New York Stock Exchange 16
New York Times 15, 38, 53, 185, 188, 276, 280
nickel 287
nitric acid 287
Nitrigin Eireann Teoranta (Ireland) 275
Nitro, West Virginia (USA) 196
nitrogen oxides 112
nitrous oxide 50–1
nitrogen 66, 302n.
Nostrom II, G. P. 211
Nuclear Regulatory Commission (USA) 239
Nypro (Britain) 275, 314n.

Oak Ridge Laboratory (USA) 299n.
Occupational Safety and Health Administration (USA) 112, 155, 160–2, 167, 175–7, 184, 203, 205

Office of Atomic Energy (Thailand) 250
Office of Management and Budget (USA) 160
Oil, Chemical and Atomic Workers' Union (USA) 161, 252-6, 280-2
Oldford, R. 169
oleum 147-8
Olin Chemical Group (USA) 193
O'Neill, L. 167
Operation Faith 63-8, 302n.
Ophthalmological Society of India 60
Organization for Economic Co-operation and Development 6, 210, 258
organo-phosphate pesticides 259
organophosphorous pesticides 314n.

Padang Jawa (Malaysia) 300n.
Paducah, Kentucky (USA) 171, 299n.
Panjwani, Mr 124
paracetic acid 203
Parkersburg, West Virginia (USA) 156
Pathak, B. 56
Patidar, S. S. 296n.
Patil, S. 73
Patil, V. 42
Pemex (Mexico) 55, 275, 309n.
People Concerned about MIC (USA) 164, 173
People's Health Clinic, see Jana Swaasthya Kendra
perchloroethylene/Perklone 288
Perrow, C. 210, 223, 243-4, 245
Personnel Research Establishment (British Army) 298n.
pesticides 211, 261, 273, 307n.
Petrobras (Brazil) 275
Petty, T. 22
phenol 265
phosgene 68, 108-9, 154, 180, 204, 225, 255-6, 259, 267, 296n.
phosphine 248
Phuket (Thailand) 247, 248-54, 272
Physicians for Social Responsibility (Switzerland) 260

Piercy, R. W. 193
pigments 287
Pinto, V. 85
plasticizers 287
polyacrylics 286
polyamides 286
polyethylene 286
polystyrene 286
polyurethanes 286
polyvinyl 286
polyvinyl chloride (PVC) 264-5, 286
Porton Down, Wiltshire (Britain) 27, 298
Prakash, P. 99, 101-2, 106
Press Trust of India 29, 51, 56
Pritchard, G. 213
propane 307n.
Public Health Engineering Department (Madhya Pradesh) 59
Purkayastha, P. 77, 274
Pyle, R. 161
pymolidone 286

Rajak, R. L. 59
Raman, A. 187
Raman, K. S. 45
Rand Corp. 125
Rao, K. M. 298n.
Rashtria Abhiyan Samiti/National Campaign Committee (India) 76, 303n.
Rasmussen, J. 223, 244
Rausch, D. A. 9, 213, 220, 242
Ray, B. 172
Red Army Faction 262
Red Cross (India) 85, 115, 127, 304n.
Red Cross (International) 127
Red Cross (USA) 127, 304n.
Rehfield, J. M. 44
Remachandran, P. K. 51
Research and Development Centre (Bhopal) 26-7, 29-30, 74, 78, 273
Robb, R. 171-2
Robinson, G. 176
rocket propellant 181

Rotterdam (Holland) 204
Roy, S. K. 82
Royal Commonwealth Society for the Blind (Britain) 23, 60
Royal Society of Chemistry (Britain) 108
runaway reactions 45, 168, 181, 183, 266, 307n.

Sadgopal, A. 71, 75, 84
Sadgopal, M. 99–101, 103, 114, 119, 298n.
Safer Emergency Systems 179
Saheli (Delhi) 75, 115
St-Louis (France) 261
Sandoz (Basle) 4, 201, 258–69, 273
Sawhney, V. 85
Schwartz, H. 262
Schwartz, V. 130
Scrip 263
Searle (USA) 202
Selikoff, I. 279
Sella, Jr, G. J. 192, 195–6
Sellafield, Cumbria (Britain) 213
Seminario, M. 155, 188
Sen, A. 128
SES Inc. 169, 180
Seveso directive 152–3, 270, 305n.
Seveso (Italy) 1, 2, 4, 17, 57, 65, 213, 222, 258, 303n., 307n.
Sevin 43, 186
Sharma, Lt-Col. 298n.
Shekhar, C. 31
Shelby, M. D. 119
Shepherd, A. J. 237
Shrader-Frachette, K. 245, 312n.
Sigg, H. P. 262
silica dust 315n.
Silicon Valley 279
Singh, A. 62, 65, 70, 96, 110, 146, 296n., 303n.
Smets, H. 210
Social Democratic Party (West Germany) 259, 266, 268
Social Welfare Department (India) 84

Society for Participatory Research in Asia 77
sodium cyanate 149
sodium hydrosulphate 147
sodium thiosulphate 106–7, 109, 117, 298n., 303n.
solvents 287–8
Somaiya Organo Chemical Co. (India) 148
South Charleston, West Virginia (USA) 165, 167, 171–4, 196, 203–5
Speth, J. 144
Spike 149
Sriram, V. 56
Sriram Food and Fertilizers Industries (Delhi) 137, 147–8
Srivastava, R. K. 298
stabilizers 287
Stafford, Senator 161
Statesman 96
Stein, R. E. 122
Story, L. J. 254, 256
Stover, W. 30, 145, 158–9, 188–90
styrene 287
Subramaniam, M. A. 77
sulphate 269
sulphur dioxide 148
Sultania Women's Hospital (Bhopal) 100
Sunday 26, 98, 111
Superfund 191
Supreme Court (India) 84, 87, 116–17, 124, 137–8, 148
Supreme Court (USA) 132
Suraksha (relief project) 85, 87
Swift, I. 182–3, 185–6
Sydney (Australia) 204
Synthetic Organic Chemical Manufacturers' Association (USA) 144

Taft, Louisiana (USA) 203
Taiwan 149
talc 288
Talty, J. T. 221–2
Tandy, H. 52
tantalum 249–52

Tata Institute of Social Sciences (India) 84
Teague, H. 113
Temik 150, 205
Teresa Mother 32, 298n.
Texaco (USA) 131
Texas City, Texas (USA) 2
Thailand 4
Thailand Tantalum Industry Corp./TTI 249–52
thiocyanate 84, 113–15
Thomas, L. M. 191, 197, 245
Thompson, E. P. 277
Three Mile Island, Pennsylvania (USA) 4, 57, 68, 144, 213, 227, 297n.
Thyagarajan, G. 309n.
Times Beach (USA) 65, 192
toluene 203
toluene di-isocyanate 255, 257–8
Toxic Catastrophe Prevention Act (New Jersey) 159
Trade Union Relief Fund (India) 76, 116
Train, R. 157, 169
Transport and General Workers' Union (Britain) 153
Transquímica (Mexico) 264
Tripathi, D. S. 82
tryamine 224
Turner & Newall (Britain) 298n.
2,4-D 265, 268
2,4,5-T 204
Tyagi, C. P. 296n.
Tylenol 32
Tyson, C. S. 33, 35
Tyson, P. 167, 176

UK Atomic Energy Authority 213
ultraviolet light 288
Union Carbide de Brasil 149–50
Union Carbide Karamchari Sangh (India) 80, 115
Union Carbide Workers' Union (Bhopal) 80
Union Research Group (Bombay) 20, 56, 75, 78, 275

United Auto Workers (USA) 155
United Nations Children's Fund 56
United Rubber Workers (USA) 253
United Steelworkers of America 161, 253
uranium 299n.
Utidjian, Dr 38

Valentine, Arizona (USA) 187
Van Mynen, R. 48, 301n.
Varadarajan, S. 35, 42, 51–2, 57, 59, 63–4, 67, 82–3, 147, 302n.
Varma, D. R. 24
Verband der Chemischen Industrie (West Germany) 267, 269
Vidal, J. 79
vinyl acetate 307n.
Voluntary Health Association of India 102
Vora, M. 69, 84
Vora, N. 79

Wages, R. E. 161
Wagoner, J. 279
Waldhut (West Germany) 265
Wall Street Journal 131, 147, 149–50, 200, 225–6, 276
Walter, I. 37, 39
Warner, F. 108
Warren County (USA) 272
Washington Post 125
Waste Insurance Liability Ltd 200
Water and Air Pollution Control Board (India) 64, 83
Waxman, H. 36, 155–5, 158–9, 175, 196, 308n.
Weill, H. 22–3
West Bengal (India) 74
West Bengal Pollution Control Board 146
West Dunbar (USA) 165
West Virginia Citizens' Action Group 165–6
West Virginia Manufacturers' Association 166

West Virginia State Air Pollution Control
    Commission 165–6, 204
West Virginia State College 165, 178
West Virginia (USA) 175, 203
White, C. 178
Wilbur-Ellis Co. (USA) 307
Wilson, D. C. 209
Winkler, H. 260
Wishart, R. 18, 135
Wolford, R. 161
Wood, K. 216
Woods, P. D. 227
Woomer, W. J. 44
World Health Organization 21, 23
World in Action 70

World Petrochemical Analysis 52
World Resources Institute 144
World Wildlife Fund (USA) 157
Wright, M. 161
Wyrick, B. 40

Yabucoa (Puerto Rico) 308n.

Zahreeli Gas Kand Sangharsh Morcha
    (India) 70–1, 74–5, 80–1, 84, 98, 105,
    115–16, 288
zinc 288
Zürich Versicherung 261–2

This first edition of
CORPORATE KILLING
Bhopals Will Happen
was finished in June 1988.

It was set in 10/14 Garamond
on a Linotron 202
and printed by a Miller TP41
on 80g/m² vol. 18 Supreme book wove.

The book was commissioned and
edited by Les Levidow,
designed by Wendy Millichap,
and produced by David Williams and
Selina O'Grady for Free Association Books.